George Darwin

On the Precession of a Viscous Spheroid and on the Remote History of the Earth

George Darwin

On the Precession of a Viscous Spheroid and on the Remote History of the Earth

ISBN/EAN: 9783337213985

Printed in Europe, USA, Canada, Australia, Japan

Cover: Foto ©berggeist007 / pixelio.de

More available books at **www.hansebooks.com**

XIII. *On the Precession of a Viscous Spheroid, and on the remote History of the Earth.*

By G. H. DARWIN, *M.A.*, *Fellow of Trinity College, Cambridge.*

Communicated by J. W. L. GLAISHER, *M.A.*, *F.R.S.*

Received July 22,—Read December 19, 1878.

[PLATE 36.]

THE following paper contains the investigation of the mass-motion of viscous and imperfectly elastic spheroids, as modified by a relative motion of their parts, produced in them by the attraction of external disturbing bodies; it must be regarded as the continuation of my previous paper,* where the theory of the bodily tides of such spheroids was given.

The problem is one of theoretical dynamics, but the subject is so large and complex, that I thought it best, in the first instance, to guide the direction of the speculation by considerations of applicability to the case of the earth, as disturbed by the sun and moon.

In order to avoid an incessant use of the conditional mood, I speak simply of the earth, sun, and moon; the first being taken as the type of the rotating body, and the two latter as types of the disturbing or tide-raising bodies. This course will be justified, if these ideas should lead (as I believe they will) to important conclusions with respect to the history of the evolution of the solar system. This plan was the more necessary, because it seemed to me impossible to attain a full comprehension of the physical meaning of the long and complex formulas which occur, without having recourse to numerical values; moreover, the differential equations to be integrated were so complex, that a laborious treatment, partly by analysis and partly by numerical quadratures, was the only method that I was able to devise. Accordingly, the earth, sun, and moon form the system from which the requisite numerical data are taken.

It will of course be understood that I do not conceive the earth to be really a homogeneous viscous or elastico-viscous spheroid, but it does seem probable that the earth still possesses some plasticity, and if at one time it was a molten mass (which is highly probable), then it seems certain that some changes in the configuration of the three bodies must have taken place, closely analogous to those hereafter determined. And even if the earth has always been quite rigid, the greater part of the same effects would result from oceanic tidal friction, although probably they would have taken place with less rapidity.

* "On the Bodily Tides of Viscous and Semi-elastic Spheroids," &c., Phil. Trans. 1879, Part I.

As some persons may wish to obtain a general idea of the drift of the inquiry without reading a long mathematical argument, I have adhered to the plan adopted in my former paper, of giving at the end (in Part III.) a general view of the whole subject, with references back to such parts as it did not seem desirable to reproduce. In order not to interrupt the mathematical argument in the body of the paper, the discussion of the physical significance of the several results is given along with the summary; such discussions will moreover be far more satisfactory when thrown into a continuous form than when scattered in isolated paragraphs throughout the paper. I have tried, however, to prevent the mathematical part from being too bald of comments, and to place the reader in a position to comprehend the general line of investigation.

Before entering on analysis, it is necessary to give an explanation of how this inquiry joins itself on to that of my previous paper.

In that paper it was shown that, if the influence of the disturbing body be expressed in the form of a potential, and if that potential be expressed as a series of solid harmonic functions of points within the disturbed spheroid, each multiplied by a simple time harmonic, then each such harmonic term raises a tide in the disturbed spheroid, which is the same as though all the other terms were non-existent. This is true, whether the spheroid be fluid, elastic, viscous, or elastico-viscous. Further, the free surface of the spheroid, as tidally distorted by any term, is expressible by a surface harmonic of the same type as that of the generating term; and where there is a frictional resistance to the tidal motion, the phase of the corresponding simple time harmonic is retarded. The height of each tide, and the retardation of phase (or the lag) are functions of the frequency of the tide, and of the constants expressive of the physical constitution of the spheroid.

Each such term in the expression for the form of the tidally distorted spheroid may be conveniently referred to as a simple tide.

Hence if we regard the whole tide-wave as a modification of the equilibrium tide-wave of a perfectly fluid spheroid, it may be said that the effect of the resistances to relative displacement is a disintegration of the whole wave into its constituent simple tides, each of which is reduced in height, and lags in time by its own special amount. In fact, the mathematical expansion in surface harmonics exactly corresponds to the physical breaking up of a single wave into a number of secondary waves.

It was remarked in the previous paper,* that when the tide-wave lags the attraction of the external tide-generating body gives rise to forces on the spheroid which are not rigorously equilibrating. Now it was a part of the assumptions, under which the theory of viscous and elastico-viscous tides was formed, that the whole forces which act on the spheroid *should* be equilibrating; but it was there stated that the couples arising from the non-equilibration of the attractions on the lagging tides were proportional to the square of the disturbing influence, and it was on this account that they were neglected in forming that theory of tides. The investigation of the effects

* "Bodily Tides," &c. Sec. 5.

which they produce in modifying the relative motion of the parts of the spheroid, that is to say in distorting the spheroid, must be reserved for a future occasion.*

The effects of these couples, in modifying the motion of the rotating spheroid as a whole, affords the subject of the present paper.

According to the ordinary theory, the tide-generating potential of the disturbing body is expressible as a series of LEGENDRE's coefficients; the term of the first order is non-existent, and the one of the second order has the type $\frac{3}{2}\cos^2-\frac{1}{2}$. Throughout this paper the potential is treated as though the term of the second order existed alone, but at the end it is shown that the term of the third order (of the type $\frac{5}{2}\cos^3-\frac{3}{2}\cos$) will have an effect which is fairly negligeable compared with that of the first term.

In order to apply the theory of elastic, viscous, and elastico-viscous tides, the first task is to express the tide-generating potential in the form of a series of solid harmonics relatively to axes fixed in the spheroid, each harmonic being multiplied by a simple time harmonic.

Afterwards it will be necessary to express that the wave surface of the distorted spheroid is the disintegration into simple lagging tides of the equilibrium tide-wave of a perfectly fluid spheroid.

The symbols expressive of the disintegration and lagging will be kept perfectly general, so that the theory will be applicable either to the assumptions of elasticity, viscosity, or elastico-viscosity, and probably to any other continuous law of resistance to relative displacement. It would not, however, be applicable to such a law as that which is *supposed* to govern the resistance to slipping of loose earth, nor to any law which assumes that there is no relative displacement of the parts of the solid, until the stresses have reached a definite magnitude.

After the form of the distorted spheroid has been found, the couples which arise from the attraction of the disturbing body on the wave surface will be found, and the rotation of the spheroid and the reaction on the disturbing body will be considered.

This preliminary explanation will, I think, make sufficiently clear the objects of the rather long introductory investigations which are necessary.

PART I.

§ 1. *The tide-generating potential.*

The disturbing body, or moon, is supposed to move in a circular orbit, with a uniform angular velocity $-\Omega$. The plane of the orbit is that of the ecliptic; for the investigation is sufficiently involved without complicating it by giving the true inclined eccentric orbit, with revolving nodes. [I hope however in a future paper to consider the secular changes in the inclination and eccentricity of the orbit and the modifications to be made in the results of the present investigation.]

* See the next paper "On Problems connected with the Tides of a Viscous Spheroid." Part I.

Let m be the moon's mass, c her distance, and $\tau=\frac{3}{2}\frac{m}{c^3}$.

Let XYZ (Plate 36, fig. 1) be rectangular axes fixed in space, XY being the ecliptic.

Let M be the moon in her orbit moving from Y towards X, with an angular velocity Ω.

Let ABC be rectangular axes fixed in the earth, AB being the equator.

Let i, ψ be the coordinates of the pole C referred to XYZ, so that i is the obliquity of the ecliptic, and $\frac{d\psi}{dt}$ the precession of the equinoxes.

Let r, θ, ϕ be the polar coordinates of any point P in the earth referred to ABC, as indicated in the figure.

Let ω_1, ω_2, ω_3 be the component angular velocities of the earth about the instantaneous positions of ABC.

Then we have, as usual, the geometrical equations,

$$\left.\begin{aligned} \frac{di}{dt} &= -\omega_1 \sin\chi + \omega_2 \cos\chi \\ \frac{d\psi}{dt}\sin i &= -\omega_1 \cos\chi - \omega_2 \sin\chi \\ -\frac{d\chi}{dt}+\frac{d\psi}{dt}\cos i &= \omega_3 \end{aligned}\right\} \quad . \quad . \quad . \quad . \quad . \quad (1)$$

Let Π cosec i be the precession of the equinoxes, or $\frac{d\psi}{dt}$, so that $\frac{d\chi}{dt}=\Pi\cot i-\omega_3$.* Now the earth rotates with a negative angular velocity, that is from B to A; therefore if we put $\frac{d\chi}{dt}=n$, n is equal to the true angular velocity of the earth $+\Pi\cot i$. But for purposes of numerical calculation n may be taken as the earth's angular velocity; and care need merely be taken that inequalities of very long period are not mistaken for secular changes.

Let the epoch be taken as the time when the colure ZC was in the plane of ZX, when χ was zero and the moon on the equator at Y. It will be convenient also to assume later that there was also an eclipse at the same instant. A number of troublesome symbols are thus got rid of, whilst the generality of the solution is unaffected.

Then by the previous definitions we have $\chi=nt$, $MN=\Omega t$, $NR=\frac{\pi}{2}-RD=\frac{\pi}{2}-(\phi-\chi)$.

Now if w be the mass of the homogeneous earth per unit volume, then the tide-generating gravitation potential V of the moon, estimated per unit volume, at the point r, θ, ϕ or P in the earth is, by the well-known formula, $V=w\tau r^2(\cos^2 PM-\frac{1}{3})$.

This is the function on which the tides depend, and as above explained, it must be

* The limit of $\Pi\cot i$ is still small when i is zero. In considering the precession with one disturbing body only, Π cosec i is merely the precession due to that body; but afterwards when the effect of the sun is added it must be taken as the full precession.

expanded in a series of solid harmonics of r, θ, ϕ, each multiplied by a simple time harmonic, which will involve n and Ω.

For brevity of notation nt, Ωt are written simply n, Ω, but wherever these symbols occur in the argument of a trigonometrical term they must be understood to be multiplied by t the time.

We have

$$\cos \mathrm{PM} = \sin\theta\cos \mathrm{MR} + \cos\theta\sin \mathrm{MR}\sin \mathrm{MRQ}$$

and

$$\cos \mathrm{MR} = \cos \mathrm{MN}\cos \mathrm{NR} + \sin \mathrm{MN}\sin \mathrm{NR}\cos i$$
$$= \cos\Omega\sin(\phi-n) + \sin\Omega\cos(\phi-n)\cos i$$

also

$$\sin \mathrm{MR}\sin \mathrm{MRQ} = \sin \mathrm{MQ} = \sin\Omega\sin i$$

Therefore

$$\cos \mathrm{PM} = \sin\theta\sin(\phi-n)\cos\Omega + \sin\theta\cos(\phi-n)\sin\Omega\cos i + \cos\theta\sin\Omega\sin i$$
$$= \tfrac{1}{2}\sin\theta\{\sin[\phi-(n-\Omega)] + \sin[\phi-(n+\Omega)]\}$$
$$+ \tfrac{1}{2}\sin\theta\cos i\{\sin[\phi-(n-\Omega)] - \sin[\phi-(n+\Omega)]\} + \cos\theta\sin\Omega\sin i$$

Let

$$p = \cos\frac{i}{2},\ q = \sin\frac{i}{2}$$

Then

$$\cos \mathrm{PM} = p^2\sin\theta\sin[\phi-(n-\Omega)] + 2pq\cos\theta\sin\Omega + q^2\sin\theta\sin[\phi-(n+\Omega)] \quad . \quad (2)$$

Therefore

$$\cos^2 \mathrm{PM} = \tfrac{1}{2}p^4\sin^2\theta\{1-\cos[2\phi-2(n-\Omega)]\} + 2p^2q^2\cos^2\theta(1-\cos 2\Omega)$$
$$+ \tfrac{1}{2}q^4\sin^2\theta\{1-\cos[2\phi-2(n+\Omega)]\} + 2p^3q\sin\theta\cos\theta\{\cos(\phi-n) - \cos[\phi-(n-2\Omega)]\}$$
$$+ 2pq^3\sin\theta\cos\theta\{\cos[\phi-(n+2\Omega)] - \cos(\phi-n)\} + p^2q^2\sin^2\theta\{\cos 2\Omega - \cos(2\phi-2n)\}$$

Then collecting terms, and noticing that

$$\tfrac{1}{2}(p^4+q^4)\sin^2\theta + 2p^2q^2\cos^2\theta = \tfrac{1}{3} + \tfrac{1}{2}(1-6p^2q^2)(\tfrac{1}{3} - \cos^2\theta)$$

we have

$$\frac{\mathrm{V}}{w\tau r^2} = \cos^2 \mathrm{PM} - \frac{1}{3}$$
$$= -\tfrac{1}{2}\sin^2\theta\{p^4\cos[2\phi-2(n-\Omega)] + 2p^2q^2\cos[2\phi-2n] + q^4\cos[2\phi-2(n+\Omega)]\}$$
$$-2\sin\theta\cos\theta\{p^3q\cos[\phi-(n-2\Omega)] - pq(p^2-q^2)\cos(\phi-n) - pq^3\cos[\phi-(n+2\Omega)]\}$$
$$+(\tfrac{1}{3}-\cos^2\theta)\{3p^2q^2\cos 2\Omega + \tfrac{1}{2}(1-6p^2q^2)\} \quad . \quad . \quad . \quad (3)$$

Now if all the cosines involving ϕ be expanded, it is clear that we have V consisting of thirteen terms which have the desired form, and a fourteenth which is independent of the time.

It will now be convenient to introduce some auxiliary functions, which may be defined thus,

$$\left.\begin{aligned}\Phi(2n)&=\tfrac{1}{2}p^4\cos 2(n-\Omega)+p^2q^2\cos 2n+\tfrac{1}{2}q^4\cos 2(n+\Omega)\\ \Psi(n)&=2p^3q\cos(n-2\Omega)-2pq(p^2-q^2)\cos n-2pq^3\cos(n+2\Omega)\\ \mathrm{X}(2\Omega)&=3p^2q^2\cos 2\Omega\end{aligned}\right\}\quad .\ .\ .\ (4)$$

$\Phi(2n-\frac{1}{2}\pi)$, $\Psi(n-\frac{1}{2}\pi)$, $\mathrm{X}(2\Omega-\frac{1}{2}\pi)$ are functions of the same form with sines replacing cosines. When the arguments of the functions are simply $2n$, n, 2Ω respectively, they will be omitted and the functions written simply Φ, Ψ, X; and when the arguments are simply $2n-\frac{1}{2}\pi$, $n-\frac{1}{2}\pi$, $2\Omega-\frac{1}{2}\pi$, they will be omitted and the functions written Φ', Ψ', X'. These functions may of course be expanded like sines and cosines, *e.g.*, $\Psi(n-\alpha)=\Psi\cos\alpha+\Psi'\sin\alpha$ and $\Psi'(n-\alpha)=\Psi'\cos\alpha-\Psi\sin\alpha$.

If now these functions are introduced into the expression for V, and if we replace the direction cosines $\sin\theta\cos\phi$, $\sin\theta\sin\phi$, $\cos\theta$ of the point P by ξ, η, ζ, we have

$$\frac{\mathrm{V}}{w\tau r^2}=-(\xi^2-\eta^2)\Phi-2\xi\eta\Phi'-\xi\zeta\Psi-\eta\zeta\Psi'+\tfrac{1}{3}(\xi^2+\eta^2-2\zeta^2)[\mathrm{X}+\tfrac{1}{2}(1-6p^2q^2)]\ .\ (5)$$

$\xi^2-\eta^2$, $2\xi\eta$, $\xi\zeta$, $\eta\zeta$, $\frac{1}{3}(\xi^2+\eta^2-2\zeta^2)$ are surface harmonics of the second order, and the auxiliary functions involve only simple harmonic functions of the time. Hence we have obtained V in the desired form.

We shall require later certain functions of the direction cosines of the moon referred to A B C expressed in terms of the auxiliary functions. The formation of these functions may be most conveniently done before proceeding further.

Let x, y, z be these direction cosines, then

$$\cos \mathrm{PM}=x\xi+y\eta+z\zeta$$

whence

$$\begin{aligned}\cos^2\mathrm{PM}-\tfrac{1}{3}&=(x\xi+y\eta+z\zeta)^2-\tfrac{1}{3}(\xi^2+\eta^2+\zeta^2)\\ &=\xi^2(x^2-\tfrac{1}{3})+\eta^2(y^2-\tfrac{1}{3})+\zeta^2(z^2-\tfrac{1}{3})+2\eta\zeta yz+2\zeta\xi zx+2\xi\eta xy\quad .\ (6)\end{aligned}$$

But from (5) we have on rearranging the terms,

$$\cos^2\mathrm{PM}-\frac{1}{3}=$$

$$\begin{aligned}&\xi^2\{-\Phi+\tfrac{1}{3}\mathrm{X}+\tfrac{1}{6}(1-6p^2q^2)\}+\eta^2\{\Phi+\tfrac{1}{3}\mathrm{X}+\tfrac{1}{6}(1-6p^2q^2)\}+\zeta^2\{-\tfrac{2}{3}\mathrm{X}-\tfrac{1}{3}(1-6p^2q^2)\}\\ &-2\eta\zeta.\tfrac{1}{2}\Psi'-2\zeta\xi.\tfrac{1}{2}\Psi-2\xi\eta\Phi'\ .\ .\ .\ .\ .\ .\ .\ .\ .\ .\ .\ .\ .\ .\ (5')\end{aligned}$$

Then equating coefficients in these two expressions (5′) and (6)

$$x^2-\tfrac{1}{3}=-\Phi+\tfrac{1}{3}X+\tfrac{1}{6}(1-6p^2q^2)$$
$$y^2-\tfrac{1}{3}=\ \ \Phi+\tfrac{1}{3}X+\tfrac{1}{6}(1-6p^2q^2)$$
$$z^2-\tfrac{1}{3}=\ \ \ -\tfrac{2}{3}X-\tfrac{1}{3}(1-6p^2q^2)$$

Whence

$$\left.\begin{aligned} y^2-z^2&=\ \ \Phi+X+\tfrac{1}{2}(1-6p^2q^2)\\ z^2-x^2&=\ \ \Phi-X-\tfrac{1}{2}(1-6p^2q^2)\\ x^2-y^2&=-2\Phi\\ \text{also}\qquad yz&=-\tfrac{1}{2}\Psi'\\ zx&=-\tfrac{1}{2}\Psi\\ xy&=-\ \Phi' \end{aligned}\right\}\qquad(7)$$

These six equations (7) are the desired functions of x, y, z in terms of the auxiliary functions.

§ 2. *The form of the spheroid as tidally distorted.*

The tide-generating potential has thirteen terms, each consisting of a solid harmonic of the second degree multiplied by a simple harmonic function of the time, viz. : three in Φ, three in Φ', three in Ψ, three in Ψ', and one in X. The fourteenth term of V can raise no proper tide, because it is independent of the time, but it produces a permanent increment to the ellipticity of the mean spheroid.

Hence according to our hypothesis, explained in the introductory remarks, there will be thirteen distinct simple tides; the three tides corresponding to Φ' may however be compounded with the three in Φ, and similarly the Ψ' tides with the Ψ tides. Hence there are seven tides with speeds* $[2n-2\Omega,\ 2n,\ 2n+2\Omega]$, $[n-2\Omega,\ n,\ n+2\Omega]$, $[2\Omega]$, and each of these will be retarded by its own special amount.

The Φ tides have periods of nearly a half-day, and will be called the slow, sidereal, and fast semi-diurnal tides, the Ψ tides have periods of nearly a day, and will be called the slow, sidereal, and fast diurnal tides, and the X tide has a period of a fortnight, and is called the fortnightly tide.

The retardation of phase of each tide will be called the "lag," and the height of each tide will be expressed as a fraction of the corresponding equilibrium tide of a perfectly fluid spheroid. Then the following schedule gives the symbols to be introduced to express lag and reduction of tide :—

* The useful term "speed" is due, I believe, to Sir WILLIAM THOMSON, and is much wanted to indicate the angular velocity of the radius of a circle, the inclination of which to a fixed radius gives the argument of a trigonometrical term. It will be used throughout this paper to indicate v, as it occurs in expressions of the type $\cos(vt+\eta)$.

	Semi-diurnal.			Diurnal.			Fortnightly.
Tide . . .	Slow ($2n-2\Omega$).	Sidereal ($2n$).	Fast ($2n+2\Omega$).	Slow ($n-2\Omega$).	Sidereal (n).	Fast ($n+2\Omega$).	(2Ω).
Height . .	E_1	E	E_2	E'_1	E'	E'_2	E''
Lag . . .	$2\epsilon_1$	2ϵ	$2\epsilon_2$	ϵ'_1	ϵ'	ϵ'_2	$2\epsilon''$

The E's are proper fractions, and the ϵ's are angles.

Let $r=a+\sigma$ be the equation to the surface of the spheroid as tidally distorted, a being the radius of the mean sphere,—for we may put out of account the permanent equatorial protuberance due to rotation, and to the non-periodic term of V.

It is a well known result that, if $wr^2\text{S}\cos(vt+\eta)$ be a tide-generating potential, estimated per unit volume of a homogeneous perfectly fluid spheroid of density w, (S being of the second order of surface harmonics), then the equilibrium tide due to this potential is given by $\sigma=\frac{5a^2}{2g}\text{S}\cos(vt+\eta)$. If we write $\mathfrak{g}=\frac{2g}{5a}$, this result may be written $\frac{\sigma}{a}=\frac{\text{S}}{\mathfrak{g}}\cos(vt+\eta)$.

Now consider a typical term—say one part of the slow semi-diurnal term—of the tide-generating potential, as found in (3): it was

$$-wr^2\tau\tfrac{1}{2}p^4\sin^2\theta\cos 2\phi\cos 2(n-\Omega).$$

The equilibrium value of the corresponding tide is found by putting $\frac{\sigma}{a}$ equal to this expression divided by $wr^2\mathfrak{g}$.

Then if we suppose that there is a frictional resistance to the tidal motion, the tide will lag and be reduced in height, and according to the preceding definitions the corresponding tide of our spheroid is expressed by

$$\frac{\sigma}{a}=-\frac{\tau}{\mathfrak{g}}E_1\tfrac{1}{2}p^4\sin^2\theta\cos 2\phi\cos\left[2(n-\Omega)-2\epsilon_1\right]$$

All the other tides may be treated in the same way, by introducing the proper E's and ϵ's.

Thus if we write

$$\left.\begin{aligned}\Phi_\epsilon&=E_1\tfrac{1}{2}p^4\cos(2n-2\Omega-2\epsilon_1)+Ep^2q^2\cos(2n-2\epsilon)+E_2\tfrac{1}{2}q^4\cos(2n+2\Omega-2\epsilon_2)\\ \Psi_\epsilon&=E'_1 2p^3q\cos(n-2\Omega-\epsilon'_1)-E'2pq(p^2-q^2)\cos(n-\epsilon')-E'_2 2pq^3\cos(n+2\Omega-\epsilon'_2)\\ \text{X}_\epsilon&=E''3p^2q^2\cos(2\Omega-2\epsilon'')\end{aligned}\right\}\quad(8)$$

and if in the same symbols accented sines replace cosines, then, by comparison with (5), we see that

$$\frac{\mathfrak{g}}{\tau}\frac{\sigma}{a}=-(\xi^2-\eta^2)\Phi_{,}-2\xi\eta\Phi'_{,}-\xi\zeta\Psi_{,}-\eta\zeta\Psi'_{,}+\tfrac{1}{3}(\xi^2+\eta^2-2\zeta^2)X_{,} \quad . \quad . \quad . \quad (9)$$

This is merely a symbolical way of writing down that every term in the tide-generating potential raises a lagging tide of its own type, but that tides of different speeds have different heights and lags.

This same expression may also be written

$$\frac{\mathfrak{g}}{\tau}\frac{\sigma}{a}=-\xi^2\{\Phi_{,}-\tfrac{1}{3}X_{,}\}-\eta^2\{-\Phi_{,}-\tfrac{1}{3}X_{,}\}-\zeta^2\tfrac{2}{3}X_{,}-2\eta\zeta\tfrac{1}{2}\Psi'_{,}-2\zeta\xi\tfrac{1}{2}\Psi_{,}-2\xi\eta\Phi'_{,} \quad . \quad . \quad . \quad (9')$$

Then if we put

$$\left.\begin{aligned}
\mathrm{c}-\mathrm{b}&=\ \ \Phi_{,}+X_{,}\\
\mathrm{a}-\mathrm{c}&=\ \ \Phi_{,}-X_{,}\\
\mathrm{b}-\mathrm{a}&=-2\Phi_{,}\\
\mathrm{c}&=\ \ \tfrac{2}{3}X_{,}\\
\mathrm{d}&=-\tfrac{1}{2}\Psi'_{,}\\
\mathrm{e}&=-\tfrac{1}{2}\Psi_{,}\\
\mathrm{f}&=-\ \Phi'_{,}
\end{aligned}\right\} \quad . \quad . \quad . \quad . \quad . \quad . \quad . \quad . \quad . \quad (10)$$

It is clear that

$$\frac{\mathfrak{g}}{\tau}\frac{\sigma}{a}=-\mathrm{a}\xi^2-\mathrm{b}\eta^2-\mathrm{c}\zeta^2+2\mathrm{d}\eta\zeta+2\mathrm{e}\zeta\xi+2\mathrm{f}\xi\eta \quad . \quad . \quad . \quad . \quad . \quad . \quad (11)$$

Whence

$$\left.\begin{aligned}
\frac{\mathfrak{g}}{2\tau}\left(\eta\frac{d}{d\zeta}-\zeta\frac{d}{d\eta}\right)\frac{\sigma}{a}&=-\{(\mathrm{c}-\mathrm{b})\eta\zeta-\mathrm{d}(\eta^2-\zeta^2)-\mathrm{e}\xi\eta+\mathrm{f}\zeta\xi\}\\
\frac{\mathfrak{g}}{2\tau}\left(\zeta\frac{d}{d\xi}-\xi\frac{d}{d\zeta}\right)\frac{\sigma}{a}&=-\{(\mathrm{a}-\mathrm{c})\zeta\xi-\mathrm{e}(\zeta^2-\xi^2)-\mathrm{f}\eta\zeta+\mathrm{d}\xi\eta\}\\
\frac{\mathfrak{g}}{2\tau}\left(\xi\frac{d}{d\eta}-\eta\frac{d}{d\xi}\right)\frac{\sigma}{a}&=-\{(\mathrm{b}-\mathrm{a})\xi\eta-\mathrm{f}(\xi^2-\eta^2)-\mathrm{d}\zeta\xi+\mathrm{e}\eta\zeta\}
\end{aligned}\right\} \quad . \quad . \quad . \quad (12)$$

Of which expressions use will be made shortly.

§ 3. *The couples about the axes* A, B, C *caused by the moon's attraction.*

The earth is supposed to be a homogeneous spheroid of mean radius a, and mass w per unit volume, so that its mass $M=\frac{4}{3}\pi wa^3$. When undisturbed by tidal distortion it is a spheroid of revolution about the axis C, and its greatest and least principal moments of inertia are C, A. Upon this mean spheroid of revolution is superposed the tide-wave σ.

The attraction of the moon on the mean spheroid produces the ordinary precessional couples $2\tau(C-A)yz$, $-2\tau(C-A)zx$, 0 about the axes A, B, C respectively; besides these there are three couples, $\mathfrak{L}$, $\mathfrak{M}$, $\mathfrak{N}$ suppose, caused by the attraction on the wave surface σ.

As it is only desired to determine the corrections to the ordinary theory of precession, the former may be omitted from consideration, and the attention confined to the determination of $\mathfrak{L}$, $\mathfrak{M}$, $\mathfrak{N}$.

The moon will be treated as an attractive particle of mass m.

Now σ as defined by (9) is a surface harmonic of the second order; hence by the ordinary formula in the theory of the potential, the gravitation potential of the tide-wave at a point whose coordinates referred to A, B, C are $r\xi$, $r\eta$, $r\zeta$ is $\frac{4}{5}\pi wa\left(\frac{a}{r}\right)^3\sigma$ or $\frac{3}{5}\frac{Ma}{r^3}\sigma$. Hence the moments about the axes A, B, C of the forces which act on a particle of mass m, situated at that point, are $\frac{3}{5}\frac{mMa}{r^3}\left(\eta\frac{d\sigma}{d\zeta}-\zeta\frac{d\sigma}{d\eta}\right)$, &c., &c. Then if this particle has the mass of the moon; if r be put equal to c, the moon's distance; and if ξ, η, ζ be replaced in σ by x, y, z (the moon's direction cosines) in the previous expressions, it is clear that $-\frac{2}{5}Ma\tau\left(y\frac{d\sigma}{dz}-z\frac{d\sigma}{dx}\right)$, &c., &c., are the couples on the earth caused by the moon's attraction.

These reactive couples are the required $\mathfrak{L}$, $\mathfrak{M}$, $\mathfrak{N}$.

Hence referring back to (12) and remarking that $\frac{2}{5}Ma^2=\mathrm{C}$, the earth's moment of inertia, we see at once that

$$\left.\begin{aligned}\frac{\mathfrak{L}}{\mathrm{C}}&=\frac{2\tau^2}{\mathfrak{g}}[(\mathrm{c}-\mathrm{b})yz-\mathrm{d}(y^2-z^2)-\mathrm{e}xy+\mathrm{f}zx]\\ \frac{\mathfrak{M}}{\mathrm{C}}&=\frac{2\tau^2}{\mathfrak{g}}[(\mathrm{a}-\mathrm{c})zx-\mathrm{e}(z^2-x^2)-\mathrm{f}yz+\mathrm{d}xy]\\ \frac{\mathfrak{N}}{\mathrm{C}}&=\frac{2\tau^2}{\mathfrak{g}}[(\mathrm{b}-\mathrm{a})xy-\mathrm{f}(x^2-y^2)-\mathrm{d}zx+\mathrm{e}yz]\end{aligned}\right\}\quad . \quad . \quad (13)$$

Where the quantities on the right-hand side are defined by the thirteen equations (7) and (10).

I shall confine my attention to determining the alteration in the uniform precession, the change in the obliquity of the ecliptic, and the tidal friction; because the nutations produced by the tidal motion will be so small as to possess no interest.

In developing $\mathfrak{L}$ and $\mathfrak{M}$ I shall only take into consideration the terms with argument n, and in $\mathfrak{N}$ only constant terms; for it will be seen, when we come to the equations of motion, that these are the only terms which can lead to the desired end.

§ 4. *Development of the couples $\mathfrak{L}$ and $\mathfrak{M}$.*

Now substitute from (7) and (10) in the first of (13), and we have

$$\frac{\mathfrak{L}}{\mathrm{C}}\div\frac{2\tau^2}{\mathfrak{g}}=-\tfrac{1}{2}\{\Phi_e+\mathrm{X}_e\}\Psi'+\tfrac{1}{2}\Psi'_e\{\Phi+\mathrm{X}+\tfrac{1}{2}(1-6p^2q^2)\}-\tfrac{1}{2}\Psi_e\Phi'+\tfrac{1}{2}\Phi'_e\Psi \quad . \quad (14)$$

A number of multiplications have now to be performed, and only those terms which contain the argument n to be retained.

The particular argument n can only arise in six ways, viz.: from products of terms with arguments $2(n-\Omega)$, $n-2\Omega$; $2n$, n; $2(n+\Omega)$, $n+2\Omega$; $n-2\Omega$, 2Ω; $n+2\Omega$, 2Ω and from terms of argument n multiplied by constant terms.

If Φ and Ψ, and Φ' and Ψ' be written underneath one another in the various combinations in which they occur in the above expression, it will be obvious that the desired argument can only arise from terms which stand one vertically over the other; this renders the multiplication easier. The Ψ, X products are comparatively easy.

Then we have

$$(\alpha)\quad -\tfrac{1}{2}\Phi_{\iota}\Psi' = -\tfrac{1}{4}[-E_1p^7q\sin(n-2\epsilon_1)+2Ep^3q^3(p^2-q^2)\sin(n-2\epsilon)+E_2pq^7\sin(n-2\epsilon_2)]$$

$$(\beta)\quad +\tfrac{1}{2}\Psi'_{\iota}\Phi = +\tfrac{1}{4}[-E'_1p^7q\sin(n+\epsilon'_1)+2E'p^3q^3(p^2-q^2)\sin(n+\epsilon')+E'_2pq^7\sin(n+\epsilon'_2)]$$

$$(\gamma)\quad -\tfrac{1}{2}\Psi_{\iota}\Phi' = \text{same as } (\beta)$$

$$(\delta)\quad +\tfrac{1}{2}\Phi'_{\iota}\Psi = \text{same as } (\alpha)$$

$$(\epsilon)\quad -\tfrac{1}{2}X_{\iota}\Psi' = -\tfrac{1}{4}[E''6p^5q^3\sin(n-2\epsilon'')-E''6p^3q^5\sin(n+2\epsilon'')]$$

$$(\zeta)\quad +\tfrac{1}{2}\Psi'_{\iota}X = +\tfrac{1}{4}[E'_16p^5q^3\sin(n-\epsilon'_1)-E'_26p^3q^5\sin(n-\epsilon'_2)]$$

$$(\eta)\quad +\tfrac{1}{4}\Psi'_{\iota}(1-6p^2q^2) = -\tfrac{1}{4}E'2pq(p^2-q^2)(1-6p^2q^2)\sin(n-\epsilon')$$

Now put $\frac{\mathfrak{L}}{C}=F\sin n+G\cos n$. Then if the expressions (α), (β) . . . (ζ) be added up when $n=\frac{\pi}{2}$, and the sum multiplied by $\frac{2\tau^2}{\mathfrak{g}}$, we shall get F; and if we perform the same addition and multiplication when $n=0$, we shall get G.

In performing the first addition the terms (α) (δ) do not combine with any other, but the terms (β), (γ), (ζ), (η) combine.

Now

$$-\tfrac{1}{2}p^7q+\tfrac{3}{2}p^5q^3=-\tfrac{1}{2}p^5q(p^2-3q^2)$$
$$p^3q^3(p^2-q^2)-\tfrac{1}{2}pq(p^2-q^2)(1-6p^2q^2)=-\tfrac{1}{2}pq(p^2-q^2)(p^4+q^4-6p^2q^2)$$
$$\tfrac{1}{2}pq^7-\tfrac{3}{2}p^3q^5=-\tfrac{1}{2}pq^5(3p^2-q^2)$$
$$-\tfrac{3}{2}p^5q^3+\tfrac{3}{2}p^3q^5=-\tfrac{3}{2}p^3q^3(p^2-q^2).$$

Hence

$$F:\frac{2\tau^2}{\mathfrak{g}}=$$
$$\tfrac{1}{2}E_1p^7q\cos 2\epsilon_1-Ep^3q^3(p^2-q^2)\cos 2\epsilon-\tfrac{1}{2}E_2pq^7\cos 2\epsilon_2$$
$$-\tfrac{1}{2}E'_1p^5q(p^2-3q^2)\cos\epsilon'_1-\tfrac{1}{2}E'pq(p^2-q^2)(p^4+q^4-6p^2q^2)\cos\epsilon'-\tfrac{1}{2}E'_2pq^5(3p^2-q^2)\cos\epsilon'_2$$
$$-\tfrac{3}{2}E''p^3q^3(p^2-q^2)\cos 2\epsilon'' \qquad (15)$$

Again for the second addition when $n=0$, we have

$$-\tfrac{1}{2}p^7q-\tfrac{3}{2}p^5q^3=-\tfrac{1}{2}p^5q(p^2+3q^2)$$
$$p^3q^3(p^2-q^2)+\tfrac{1}{2}pq(p^2-q^2)(1-6p^2q^2)=\tfrac{1}{2}pq(p^2-q^2)^3$$
$$\tfrac{1}{2}pq^7+\tfrac{3}{2}p^3q^5=\tfrac{1}{2}pq^5(3p^2+q^2)$$
$$\tfrac{3}{2}p^5q^3+\tfrac{3}{2}p^3q^5=\tfrac{3}{2}p^3q^3,$$

So that

$$\begin{aligned}\mathrm{G}\div\frac{2\tau^2}{\mathfrak{g}}=&-\tfrac{1}{2}E_1p^7q\sin 2\epsilon_1+Ep^3q^3(p^2-q^2)\sin 2\epsilon+\tfrac{1}{2}E_2pq^7\sin 2\epsilon_2\\ &-\tfrac{1}{2}E'_1p^5q(p^2+3q^2)\sin\epsilon'_1+\tfrac{1}{2}E'pq(p^2-q^2)^3\sin\epsilon'+\tfrac{1}{2}E'_2pq^5(3p^2+q^2)\sin\epsilon'_2\\ &+\tfrac{3}{2}E''p^3q^3\sin 2\epsilon'' \quad . \quad . \quad . \quad (16)\end{aligned}$$

And

$$\frac{\mathfrak{L}}{\mathrm{C}}=\mathrm{F}\sin n+\mathrm{G}\cos n \quad . \quad . \quad . \quad (17)$$

To find M it is only necessary to substitute $n-\frac{\pi}{2}$ for n, and we have

$$\frac{\mathfrak{M}}{\mathrm{C}}=-\mathrm{F}\cos n+\mathrm{G}\sin n \quad . \quad (18)$$

Now there is a certain approximation which gives very nearly correct results and which simplifies these expressions very much. It has already been remarked that the three Φ-tides have periods of nearly a half-day and the three Ψ-tides of nearly a day, and this will continue to be true so long as Ω is small compared with n; hence it may be assumed with but slight error that the semi-diurnal tides are all retarded by the same amount and that their heights are proportional to the corresponding terms in the tide-generating potential. That is, we may put $\epsilon_1=\epsilon_2=\epsilon$ and $E_1=E_2=E$. The similar argument with respect to the diurnal tides permits us to put $\epsilon'_1=\epsilon'_2=\epsilon'$ and $E'_1=E'_2=E'$.

Then introducing the quantities $P=p^2-q^2=\cos i$, $Q=2pq=\sin i$ and observing that

$$\tfrac{1}{2}p^7q-p^3q^3(p^2-q^2)-\tfrac{1}{2}pq^7=\tfrac{1}{2}pq[(p^2-q^2)(p^4+p^2q^2+q^4)-2p^2q^2(p^2-q^2)]=\tfrac{1}{4}PQ(1-\tfrac{3}{4}Q^2)$$
$$\begin{aligned}\tfrac{1}{2}p^5q(p^2-3q^2)+\tfrac{1}{2}pq(p^2-q^2)(p^4+q^4-6p^2q^2)+\tfrac{1}{2}pq^5(3p^2-q^2)&=pq(p^2-q^2)(1-6p^2q^2)\\&=\tfrac{1}{2}PQ(1-\tfrac{3}{2}Q^2)\end{aligned}$$
$$\tfrac{1}{2}p^5q(p^2+3q^2)-\tfrac{1}{2}pq(p^2-q^2)^3-\tfrac{1}{2}pq^5(3p^2+q^2)=\tfrac{1}{2}pq(p^2-q^2)(1+2p^2q^2-1+4p^2q^2)=\tfrac{3}{8}PQ^3$$

we have,

$$\left.\begin{aligned} \mathrm{F}\div\frac{\tau^2}{\mathfrak{g}}&=\tfrac{1}{2}EPQ(1-\tfrac{3}{4}Q^2)\cos 2\epsilon-E'PQ(1-\tfrac{3}{2}Q^2)\cos\epsilon'-\tfrac{3}{8}E''PQ^3\cos 2\epsilon'' \\ \mathrm{G}\div\frac{\tau^2}{\mathfrak{g}}&=-\tfrac{1}{2}EPQ(1-\tfrac{3}{4}Q^2)\sin 2\epsilon-\tfrac{3}{4}E'PQ^3\sin\epsilon'+\tfrac{3}{8}E''Q^3\sin 2\epsilon'' \end{aligned}\right\} \quad (19)$$

§ 5. *Development of the couple* $\mathfrak{N}$.

In the couple $\mathfrak{N}$ about the axis of rotation of the earth we only wish to retain non-periodic terms, and these can only arise from the products of terms with the same argument.

By substitution from (7) and (10) in the last of (13)

$$\frac{\mathfrak{N}}{\mathrm{C}}\div\frac{2\tau^2}{\mathfrak{g}}=2\Phi_{\iota}\Phi'-2\Phi'_{\iota}\Phi-\tfrac{1}{4}\Psi'_{\iota}\Psi+\tfrac{1}{4}\Psi_{\iota}\Psi' \quad . \quad . \quad . \quad . \quad . \quad . \quad (20)$$

Then as far as we are now interested,

$$2\Phi_{\iota}\Phi'=-2\Phi'_{\iota}\Phi=E_1\tfrac{1}{4}p^8\sin 2\epsilon_1+Ep^4q^4\sin 2\epsilon+E_2\tfrac{1}{4}q^8\sin 2\epsilon_2$$
$$-\tfrac{1}{4}\Psi'_{\iota}\Psi=\tfrac{1}{4}\Psi_{\iota}\Psi'=E'_1\tfrac{1}{2}p^6q^2\sin\epsilon'_1+E'\tfrac{1}{2}p^2q^2(p^2-q^2)^2\sin\epsilon'+E'_2\tfrac{1}{2}p^2q^6\sin\epsilon'_2$$

Hence

$$\begin{aligned}\frac{\mathfrak{N}}{\mathrm{C}}\div\frac{\tau^2}{\mathfrak{g}}=&E_1p^8\sin 2\epsilon_1+E4p^4q^4\sin 2\epsilon+E_2q^8\sin 2\epsilon_2 \\ &+E'_12p^6q^2\sin\epsilon'_1+E'2p^2q^2(p^2-q^2)^2\sin\epsilon'+E'_22p^2q^6\sin\epsilon'_2 \quad . \quad . \quad . \quad (21)\end{aligned}$$

If as in the last section we group the semi-diurnal and diurnal terms together and put $E_1=E_2=E$, &c., and observe that

$$p^8+4p^4q^4+q^8=(p^4+q^4)^2+2p^4q^4=(1-\tfrac{1}{2}Q^2)^2+\tfrac{1}{8}Q^4=P^2+\tfrac{3}{8}Q^4$$
$$2p^6q^2+2p^2q^2(p^2-q^2)^2+2p^2q^6=4p^2q^2[p^4+q^4-p^2q^2]=Q^2(1-\tfrac{3}{4}Q^2),$$

then

$$\frac{\mathfrak{N}}{\mathrm{C}}\div\frac{\tau^2}{\mathfrak{g}}=E(P^2+\tfrac{3}{8}Q^4)\sin 2\epsilon+E'Q^2(1-\tfrac{3}{4}Q^2)\sin\epsilon' \quad (22)$$

§ 6. *The equations of motion of the earth abouts its centre of inertia.*

In forming the equations of motion we are met by a difficulty, because the axes A, B, C are neither principal axes, nor can they rigorously be said to be fixed in the earth. But M. LIOUVILLE has given the equations of motion of a body which is changing its shape, using any set of rectangular axes which move in any way with reference to the body, except that the origin always remains at the centre of inertia.

If A, B, C, D, E, F be the moments and products of inertia of the body about these axes of reference at any time ; H_1, H_2, H_3 the moments of momentum of the motion of all the parts of the body relative to the axes ; ω_1, ω_2, ω_3 the component angular velocities of the axes about their instantaneous positions, the equations may be written

$$\frac{d}{dt}(A\omega_1-F\omega_2-E\omega_3+H_1)+D(\omega^2_3-\omega^2_2)+(C-B)\omega_2\omega_3+F\omega_3\omega_1-E\omega_2\omega_1$$
$$+\omega_2 H_3-\omega_3 H_2=L \quad . \quad . \quad . \quad . \quad . \quad . \quad . \quad . \quad . \quad . \quad (23)$$

and two other equations found from this by cyclical changes of letters and suffixes.*

Now in the case to be considered here the axes A, B, C always occupy the average position of the same line of particles, and they move with very nearly an ordinary uniform precessional motion. Also the moments and products of inertia may be written A+a′, B+b′, C+c′, d′, e′, f′, where a′, b′, c′, d′, e′, f′ are small periodic functions of the time and a′+b′+c′=0, and where A, B, C are the principal moments of inertia of the undisturbed earth, so that B is equal to A.

Now the quantities a′, b′, &c., have in effect been already determined, as may be shown as follows: By the ordinary formula† the force function of the moon's action on the earth is $\frac{Mm}{c}+\tau\left(\frac{A+B+C}{3}-I\right)$, where I is the moment of inertia of the earth about the line joining its centre to the moon, and is therefore

$$=Ax^2+By^2+Cz^2+a'x^2+b'y^2+c'z^2-2d'yz-2e'zx-2f'xy.$$

But the first three terms of I only give rise to the ordinary precessional couples, and a comparison of the last six with (11) and (13) shows that

$$\frac{a'}{a}=\frac{b'}{b}=\frac{c'}{c}=\frac{d'}{d}=\frac{e'}{e}=\frac{f'}{f}=\frac{\tau}{g}.C.$$

Also in the small terms we may ascribe to ω_1, ω_2, ω_3 their uniform precessional values, viz. : $\omega_1=-\Pi\cos n$, $\omega_2=-\Pi\sin n$, $\omega_3=-n$.

When these values are substituted in (23), we get some small terms of the form $a'\Pi^2\sin n$, and others of the form $a'\Pi n\sin n$; both these are very small compared to the terms in $\mathfrak{L}$ and $\mathfrak{M}$—the fractions which express their relative magnitude being $\frac{\Pi^2}{\tau}$ and $\frac{\Pi n}{\tau}$.

There is also a term $-\Pi H_3\sin n$, which I conceive may also be safely neglected, as also the similar terms in the second and third equations.

It is easy, moreover, to show that according to the theories of the tidal motion of a homogeneous viscous spheroid given in the previous paper, and according to

* ROUTH's 'Rigid Dynamics' (first edition only), p. 150, or my paper in the Phil. Trans. 1877, Vol. 167, p. 272. The original is in LIOUVILLE's Journal, 2nd series, vol. iii., 1858, p. 1.

† ROUTH's 'Rigid Dynamics,' 1877, p. 495.

Sir WILLIAM THOMSON's theory of elastic tides, H_1, H_2, H_3 are all zero. Those theories both neglect inertia but the actuality is not likely to differ materially therefrom.

Thus every term where ω_1 and ω_2 occur may be omitted and the equations reduced to

$$\left.\begin{aligned} A\frac{d\omega_1}{dt}+(C-B)\omega_2\omega_3+n\frac{de'}{dt}+n^2d'+\frac{dH_1}{dt}+nH_2&=\mathfrak{L} \\ B\frac{d\omega_2}{dt}+(A-C)\omega_3\omega_1+n\frac{dd'}{dt}-n^2e'+\frac{dH_2}{dt}-nH_1&=\mathfrak{M} \\ C\frac{d\omega_3}{dt}+(B-A)\omega_1\omega_2\qquad\qquad\qquad+\frac{dH_3}{dt}\qquad&=\mathfrak{N} \end{aligned}\right\} \quad . \quad . \quad . \quad (24)$$

As before with the couples, so here, we are only interested in terms with the argument n in the small terms on the left-hand side of the first two of equations (24), and in non-periodic terms in the last of them.

Now for each term in the moon's potential, as developed in Section 1, there is (by hypothesis) a corresponding co-periodic flux and reflux throughout the earth's mass, and therefore the H_1, H_2, H_3 must each have periodic terms corresponding to each term in the moon's potential. Hence the only term in the moon's potential to be considered is that with argument n, with respect to H_1 and H_2 in the first two equations; and H_3 may be omitted from the third as being periodic.

Suppose then that H_1 was equal to $h \cos n+h' \sin n$, then precisely as we found $\mathfrak{M}$ from $\mathfrak{L}$ by writing $n-\frac{\pi}{2}$ for n we have $H_2=h \sin n-h' \cos n$. Thus $\frac{dH_1}{dt}+nH_2=0$, $\frac{dH_2}{dt}-nH_1=0$, and the H's disappear from the first two equations.

Next retaining only terms in argument n in d′ and e′, we have from (10)

$$e'=C\frac{\tau}{g}E'pq(p^2-q^2)\cos(n-\epsilon'),\quad d'=C\frac{\tau}{g}E'pq(p^2-q^2)\sin(n-\epsilon')$$

Therefore $\frac{de'}{dt}+nd'=0$, $\frac{dd'}{dt}-ne'=0$, and these terms also disappear.

Lastly, put B=A, and our equations reduce simply to those of EULER, viz.:

$$\left.\begin{aligned} A\frac{d\omega_1}{dt}+(C-A)\omega_2\omega_3&=\mathfrak{L} \\ A\frac{d\omega_2}{dt}-(C-A)\omega_3\omega_1&=\mathfrak{M} \\ C\frac{d\omega_3}{dt}\qquad\qquad\qquad&=\mathfrak{N} \end{aligned}\right\} \quad . \quad . \quad . \quad . \quad . \quad . \quad . \quad . \quad (25)$$

Now $\mathfrak{N}$ is small, and therefore ω_3 remains approximately constant and equal to $-n$ for long periods, and as C−A is small compared to A, we may put $\omega_3=-n$ in the first two equations. But when C−A is neglected compared to C, the integrals of these equations are the same as those of

$$\frac{d\omega_1}{dt}=\frac{\mathfrak{L}}{C},\quad \frac{d\omega_2}{dt}=\frac{\mathfrak{M}}{C},\quad \frac{d\omega_3}{dt}=\frac{\mathfrak{N}}{C} \quad . \quad . \quad . \quad . \quad . \quad . \quad . \quad . \quad . \qquad (26)$$

apart from the complementary function, which may obviously be omitted. The two former of (26) give the change in the precession and the obliquity of the ecliptic, and the last gives the tidal friction.

§ 7. *Precession and change of obliquity.*

Then by (17), (18), and (26) the equations of motion are

$$\left.\begin{aligned}\frac{d\omega_1}{dt}&=\ \ \mathrm{F}\sin n+\mathrm{G}\cos n\\ \frac{d\omega_2}{dt}&=-\mathrm{F}\cos n+\mathrm{G}\sin n\end{aligned}\right\} \quad . \qquad (27)$$

and by integration

$$\omega_1=\frac{1}{n}[-\mathrm{F}\cos n+\mathrm{G}\sin n],\quad \omega_2=\frac{1}{n}[-\mathrm{F}\sin n-\mathrm{G}\cos n] \quad . \quad . \quad . \quad . \qquad (28)$$

But the geometrical equations (1) give

$$\frac{di}{dt}=-\omega_1\sin n+\omega_2\cos n$$

$$\frac{d\psi}{dt}\sin i=-\omega_1\cos n-\omega_2\sin n$$

Therefore, as far as concerns non-periodic terms,

$$\frac{di}{dt}=-\frac{\mathrm{G}}{n},\quad \frac{d\psi}{dt}\sin i=\frac{\mathrm{F}}{n} \quad . \quad . \quad . \quad . \quad . \quad . \quad . \quad . \quad . \qquad (29)$$

If we wish to keep all the seven tides distinct (as will have to be done later), we may write down the result for $\frac{di}{dt}$ and $\frac{d\psi}{dt}$ from (15) and (16).

But it is of more immediate interest to consider the case where the semi-diurnal tides are grouped together, as also the diurnal ones. In this case we have by (19)

$$\frac{di}{dt}=\frac{\tau^2}{\mathfrak{g}n}\{\tfrac{1}{2}PQ(1-\tfrac{3}{4}Q^2)E\sin 2\epsilon+\tfrac{3}{4}PQ^3E'\sin\epsilon'-\tfrac{3}{8}Q^3E''\sin 2\epsilon''\} \quad . \quad . \quad . \qquad (30)$$

and since $\sin i=Q$

$$\frac{d\psi}{dt}=\frac{\tau^2}{\mathfrak{g}n}\{\tfrac{1}{2}P(1-\tfrac{3}{4}Q^2)E\cos 2\epsilon-P(1-\tfrac{3}{2}Q^2)E'\cos\epsilon'-\tfrac{3}{8}PQ^2E''\cos 2\epsilon''\} \quad . \qquad (31)$$

In these equations P and Q stand for the cosine and sine of the obliquity of the ecliptic.

Several conclusions may be drawn from this result.

If ϵ, ϵ', ϵ'' are zero the obliquity remains constant.

Now if the spheroid be perfectly elastic, the tides do not lag, and therefore the obliquity remains unchanged; it would also be easy to find the correction to the precession to be applied in the case of elasticity.

It is possible that the investigation is not, strictly speaking, applicable to the case of a perfect fluid; I shall, however, show to what results it leads if we make the application to that case. Sir WILLIAM THOMSON has shown that the period of free vibration of a fluid sphere of the density of the earth would be about 1 hour 34 minutes.* And as this free period is pretty small compared to the forced period of the tidal oscillation, it follows that E, E', E'', will not differ much from unity. Then putting them equal to unity, and putting ϵ, ϵ', ϵ'' zero, since the tides do not lag, we find that the obliquity remains constant, and

$$\frac{d\psi}{dt}=-\frac{\tau^2}{\mathfrak{g}n}\tfrac{1}{2}P(1-\tfrac{3}{2}Q^2)=-\tfrac{1}{2}\frac{\tau^2}{\mathfrak{g}n}\cos i(1-\tfrac{3}{2}\sin^2 i) \quad . \quad . \quad . \quad . \quad (32)$$

This equation gives the correction to be applied to the precession as derived from the assumption that the rotating spheroid of fluid is rigid. This result is equally true if all the seven tides are kept distinct. Now if the spheroid were rigid its precession would be $\frac{\tau e}{n}\cos i$, where e is the ellipticity of the spheroid.

The ellipticity of a fluid spheroid rotating with an angular velocity n is $\frac{5}{4}\frac{n^2a}{g}$ or $\frac{1}{2}\frac{n^2}{\mathfrak{g}}$; but besides this, there is ellipticity due to the non-periodic part of the tide-generating potential.

By (3) § 1 the non-periodic part of V is $\frac{1}{2}w\tau r^2(\frac{1}{3}-\cos^2\theta)(1-6p^2q^2)$; such a disturbing potential will clearly produce an ellipticity $\frac{1}{2}\frac{\tau}{\mathfrak{g}}(1-6p^2q^2)$.

If therefore we put $e_0=\frac{1}{2}\frac{n^2}{\mathfrak{g}}$, and remember that $6p^2q^2=\frac{3}{2}\sin^2 i$, we have,

$$e=e_0+\tfrac{1}{2}\frac{\tau}{\mathfrak{g}}(1-\tfrac{3}{2}\sin^2 i)$$

Hence if the spheroid were rigid, and had its actual ellipticity, we should have

$$\frac{d\psi}{dt}=\frac{\tau e_0}{n}\cos i+\tfrac{1}{2}\frac{\tau^2}{\mathfrak{g}n}\cos i(1-\tfrac{3}{2}\sin^2 i). \quad . \quad . \quad . \quad . \quad . \quad (32')$$

* Phil. Trans., 1863, p. 608.

Adding (32′) to (32), the whole precession is

$$\frac{d\psi}{dt}=\frac{\tau e_0}{n}\cos i \tag{32''}$$

We thus see that the effect of the non-periodic part of the tide-generating potential, which may be conveniently called a permanent tide, is just such as to neutralise the effects of the tidal action. The result (32″) may be expressed as follows:—

The precession of a fluid spheroid is the same as that of a rigid one which has an ellipticity equal to that due to the rotation of the spheroid.

From this it follows that the precession of a fluid spheroid will differ by little from that of a rigid one of the same ellipticity, if the additional ellipticity due to the non-periodic part of the tide-generating influence is small compared with the whole ellipticity.

Sir WILLIAM THOMSON has already expressed himself to somewhat the same effect in an address to the British Association at Glasgow.*

Since $e_0=\frac{1}{2}\frac{n^2}{\mathfrak{g}}$, the criterion is the smallness of $\frac{\tau}{n^2}$.

It may be expressed in a different form; for $\frac{\tau}{n^2}$ is small when $\frac{\tau e}{n}\div n$ is small compared with e, and $\frac{\tau e}{n}\div n$ is the reciprocal of the precessional period expressed in days. Hence the criterion may be stated thus: *The precession of a fluid spheroid differs by little from that of a rigid one of the same ellipticity, when the precessional period of the spheroid expressed in terms of its rotation is large compared with the reciprocal of its ellipticity.*

In his address, Sir WILLIAM THOMSON did not give a criterion for the case of a fluid spheroid without any confining shell, but for the case of a thin rigid spheroidal shell enclosing fluid he gave a statement which involves the above criterion, save that the ellipticity referred to is that of the shell itself; for he says, "The amount of this difference (in precession and nutation) bears the same proportion to the actual precession or nutation as the fraction measuring the periodic speed of the disturbance (in terms of the period of rotation as unity) bears to the fraction measuring the interior ellipticity of the shell."

This is, in fact, almost the same result as mine.

This subject is again referred to in Part III. of the succeeding paper.

* See 'Nature,' September 14, 1876, p. 429. The above statement of results, and the comparison with Sir WILLIAM THOMSON's criterion was added to the paper on September 17, 1879.

§ 8. *The disturbing action of the sun.*

Now suppose that there is a second disturbing body, which may be conveniently called the sun.*

* It is not at first sight obvious how it is physically possible that the sun should exercise an influence on the moon-tide, and the moon on the sun-tide, so as to produce a secular change in the obliquity of the ecliptic and to cause tidal friction, for the periods of the sun and moon about the earth are different. It seems, therefore, interesting to give a physical meaning to the expansion of the tide-generating potential; it will then be seen that the interaction with which we are here dealing must occur.

The expansion of the potential given in Section 1 is equivalent to the following statement:—

The tide-generating potential of a moon of mass m, moving in a circular orbit of obliquity i at a distance c, is equal to the tide-generating potential of ten satellites at the same distance, whose orbits, masses, and angular velocities are as follows:—

1. A satellite of mass $m\cos^4\frac{i}{2}$, moving in the equator in the same direction and with the same angular velocity as the moon, and coincident with it at the nodes. This gives the slow semi-diurnal tide of speed $2(n-\Omega)$.

2. A satellite of mass $m\sin^4\frac{i}{2}$, moving in the equator in the opposite direction from that of the moon, but with the same angular velocity, and coincident with it at the nodes. This gives the fast semi-diurnal tide of speed $2(n+\Omega)$.

3. A satellite of mass $m\,2\sin^2\frac{i}{2}\cos^2\frac{i}{2}$, fixed at the moon's node. This gives the sidereal semi-diurnal tide of speed $2n$.

4. A repulsive satellite of mass $-m.2\sin\frac{i}{2}\cos^3\frac{i}{2}$, moving in N. declination 45° with twice the moon's angular velocity, in the same direction as the moon, and on the colure 90° in advance of the moon, when she is in her node.

5. A satellite of mass $m\sin i\cos^3\frac{i}{2}$, moving in the equator with twice the moon's angular velocity, and in the same direction, and always on the same meridian as the fourth satellite. (4) and (5) give the slow diurnal tide of speed $n-2\Omega$.

6. A satellite of mass $m\sin^3\frac{i}{2}\cos\frac{i}{2}$, moving in N. declination 45° with twice the moon's angular velocity, but in the opposite direction, and on the colure 90° in advance of the moon when she is in her node.

7. A repulsive satellite of mass $-m.\frac{1}{2}\sin^3\frac{i}{2}\cos\frac{i}{2}$, moving in the equator with twice the moon's angular velocity, but in the opposite direction, and always on the same meridian as the sixth satellite. (6) and (7) give the fast semi-diurnal tide of $n+2\Omega$.

8. A satellite of mass $m\sin i\cos i$ fixed in N. declination 45° on the colure.

9. A repulsive satellite of mass $-m.\frac{1}{2}\sin i\cos i$, fixed in the equator on the same meridian as the eighth satellite. (8) and (9) give the sidereal diurnal tide of speed n.

10. A ring of matter of mass m, always passing through the moon and always parallel to the equator. This ring, of course, executes a simple harmonic motion in declination, and its mean position is the equator. This gives the fortnightly tide of speed 2Ω.

Now if we form the potentials of each of these satellites, and omit those parts which, being independent of the time, are incapable of raising tides, and add them altogether, we shall obtain the expansion for the moon's tide-generating potential used above; hence this system of satellites is mechanically

Π cosec i must henceforth be taken as the full precession of the earth, and the time may be conveniently measured from an eclipse of the sun or moon. Let $m_{,}$, $c_{,}$ be the sun's mass and distance; $\Omega_{,}$ the earth's angular velocity in a circular orbit; and let $\tau_{,}=\frac{3}{2}\frac{m_{,}}{c_{,}^{3}}$.

It would be rigorously necessary to introduce a new set of quantities to give the heights and lagging of the seven solar tides: but of the three solar semi-diurnal tides, one has rigorously the same period as one of the three lunar semi-diurnal tides (viz.: the sidereal semi-diurnal with a speed $2n$), and the others have nearly the same period; a similar remark applies to the solar diurnal tides. Hence we may, without much error, treat E, ϵ, E', ϵ' as the same both for lunar and solar tides; but E''', ϵ''' must replace E'', ϵ'', because the semi-annual replaces the fortnightly tide.

Then if new auxiliary functions $\Phi_{,}$, $\Psi_{,}$, $X_{,}$ be introduced, the whole tide-generating potential V per unit volume of the earth at the point $r\xi$, $r\eta$, $r\zeta$ is given by

$$\frac{V}{wr^2}=-(\tau\Phi+\tau_{,}\Phi_{,})(\xi^2-\eta^2) \text{ \&c.}$$

If then, as in (10), we put

$$c-b=\Phi_{,}+X_{,}, \text{ \&c.}, \quad c_{,}-b_{,}=\Phi_{,,}+X_{,,}, \text{ \&c.},$$

the equation to the tidally-distorted earth is $r=a+\sigma+\sigma_{,}$, where

equivalent to the action of the moon alone. The satellites 1, 2, 3, in fact, give the semi-diurnal or Φ terms; satellites 4, 5, 6, 7, 8, 9 give the diurnal or Ψ terms; and satellite 10 gives the fortnightly or X term.

This is analogous to "GAUSS's way of stating the circumstances on which 'secular' variations in the elements of the solar system depend;" and the analysis was suggested to me by a passage in THOMSON and TAIT's 'Nat. Phil.,' § 809, who there refer to the annular satellite 10.

It will appear in Section 22 that the 3rd, 8th, and 9th satellites, which are fixed in the heavens and which give the sidereal tides, are equivalent to a distribution of the moon's mass in the form of a uniform circular ring coincident with her orbit. And perhaps some other simpler plan might be given which would replace the other repulsive satellites.

These tides, here called "sidereal," are known, in the reports of the British Association on tides for 1872 and 1876, as the K tides.

In a precisely similar way, it is clear that the sun's influence may be analysed into the influence of nine other satellites and one ring, or else to seven satellites and two rings. Then, with regard to the interaction of sun and moon, it is clear that those satellites of each system which are fixed in each system (viz.: 3, 8, and 9), or their equivalent rings, will not only exercise an influence on the tides raised by themselves, but each will necessarily exercise an influence on the tides raised by the other, so as to produce tidal friction. All the other satellites will, of course, attract or repel the tides of all the other satellites of the other systems; but this interaction will necessarily be periodic, and will not cause any interaction in the way of tidal friction or change of obliquity, and as such periodic interaction is of no interest in the present investigation it may be omitted from consideration. In the analysis of the present section, this omission of all but the fixed satellites appears in the form of the omission of all terms involving the moon's or sun's angular velocity round the earth.

$$\frac{\mathfrak{g}}{\tau}\frac{\sigma}{a}=-\mathrm{a}\xi^2 \text{ \&c., } \frac{\mathfrak{g}}{\tau_{\prime}}\frac{\sigma_{\prime}}{a}=-\mathrm{a}_{\prime}\xi^2, \text{ \&c.}$$

Also if x, y, z and $x_{\prime}, y_{\prime}, z_{\prime}$ be the moon's and sun's direction cosines, we have as in (7),

$$y^2-z^2=\Phi+\mathrm{X}+\tfrac{1}{2}(1-6p^2q^2), \text{ \&c., } y_{\prime}^2-z_{\prime}^2=\Phi_{\prime}+\mathrm{X}_{\prime}+\tfrac{1}{2}(1-6p^2q^2), \text{ \&c.}$$

Then using the same arguments as in Section 3, the couples about the three axes in the earth may be found, and we have

$$\frac{\mathfrak{L}}{\mathrm{C}}=-\left\{\tau\left(y\frac{d}{dz}-z\frac{d}{dy}\right)\left(\frac{\sigma}{a}+\frac{\sigma_{\prime}}{a}\right)+\tau_{\prime}\left(y_{\prime}\frac{d}{dz_{\prime}}-z_{\prime}\frac{d}{dy_{\prime}}\right)\left(\frac{\sigma}{a}+\frac{\sigma_{\prime}}{a}\right)\right\}$$

where in the first term x, y, z are written for ξ, η, ζ in $\sigma+\sigma_{\prime}$, and in the second term $x_{\prime}, y_{\prime}, z_{\prime}$ are similarly written for ξ, η, ζ.

Now let $\mathfrak{L}_{m^2}$, $\mathfrak{L}_{m_{\prime}^2}$, $\mathfrak{L}_{mm_{\prime}}$ indicate the parts of the couple $\mathfrak{L}$ which depend on the moon's action on the lunar tides, the sun's action on the solar tides, and the moon's and sun's action on the solar and lunar tides respectively, then

$$\frac{\mathfrak{L}}{\mathrm{C}}=\frac{\mathfrak{L}_{m^2}}{\mathrm{C}}+\frac{\mathfrak{L}_{m_{\prime}^2}}{\mathrm{C}}-\left\{\tau\left(y\frac{d}{dz}-z\frac{d}{dy}\right)\frac{\sigma_{\prime}}{a}+\tau_{\prime}\left(y_{\prime}\frac{d}{dz_{\prime}}-z_{\prime}\frac{d}{dy_{\prime}}\right)\frac{\sigma}{a}\right\}$$

Then obviously

$$\frac{\mathfrak{L}_{mm_{\prime}}}{\mathrm{C}}\div\frac{2\tau\tau_{\prime}}{\mathfrak{g}}=(\mathrm{c}-\mathrm{b})y_{\prime}z_{\prime}+(\mathrm{c}_{\prime}-\mathrm{b}_{\prime})yz+\text{ \&c.}$$

As before, we only want terms with argument n in $\mathfrak{L}_{mm_{\prime}}$, $\mathfrak{M}_{mm_{\prime}}$, and non-periodic terms in $\mathfrak{N}_{mm_{\prime}}$.

The quantities a, b, &c., x, y, z with suffixes differ from those without in having $\Omega_{\prime}$ in place of Ω, and it is clear that no combination of terms which involve $\Omega_{\prime}$ and Ω can give the desired terms in the couples. Hence, as far as $\mathfrak{L}_{mm_{\prime}}$, $\mathfrak{M}_{mm_{\prime}}$, $\mathfrak{N}_{mm_{\prime}}$ are concerned, the auxiliary functions may be abridged by the omission of all terms involving Ω or $\Omega_{\prime}$.

Therefore, from (4), we now simply have

$$\Phi=\Phi_{\prime}=p^2q^2\cos 2n,\ \Psi=\Psi_{\prime}=-2pq(p^2-q^2)\cos n,\ \mathrm{X}=\mathrm{X}_{\prime}=0.$$

But $\mathrm{c}-\mathrm{b}$ only differs from $\mathrm{c}_{\prime}-\mathrm{b}_{\prime}$ in that the latter involves $\Omega_{\prime}$ instead of Ω, and the same applies to yz and $y_{\prime}z_{\prime}$.

Hence, as far as we are now concerned,

$$(\mathrm{c}-\mathrm{b})y_{\prime}z_{\prime}=(\mathrm{c}_{\prime}-\mathrm{b}_{\prime})yz$$

and similarly each pair of terms in $\mathfrak{L}_{mm_{\prime}}$ are equal *inter se.*

Thus

$$\frac{\mathfrak{L}_{mm_{\prime}}}{C}\div\frac{4\tau\tau_{\prime}}{\mathfrak{g}}=(c-b)yz-d(y^2-z^2)-exy+fzx.$$

Comparing with (14), when X is put equal to zero, we have

$$\frac{\mathfrak{L}_{mm_{\prime}}}{C}\div\frac{4\tau\tau_{\prime}}{\mathfrak{g}}=-\tfrac{1}{2}\Phi_{\prime}\Psi'+\tfrac{1}{2}\Psi'_{\prime}\{\Phi+\tfrac{1}{2}(1-6p^2q^2)\}-\tfrac{1}{2}\Psi_{\prime}\Phi'+\tfrac{1}{2}\Phi'_{\prime}\Psi.$$

This quantity may be evaluated at once by reference to (15), (16), and (17), for it is clear that $\mathfrak{L}_{mm_{\prime}}$ is what $\mathfrak{L}_{m^2}$ becomes when $E_1=E_2=0$, $E'_1=E'_2=0$, and when $2\tau\tau_{\prime}$ replaces τ^2.

If, therefore, we put $\frac{\mathfrak{L}_{mm_{\prime}}}{C}=F_{mm_{\prime}}\sin n+G_{mm_{\prime}}\cos n$, and remark that

$$4p^3q^3(p^2-q^2)=\tfrac{1}{2}PQ^3,\ 2pq(p^2-q^2)(p^4+q^4-6p^2q^2)=PQ(1-2Q^2),\ 2pq(p^2-q^2)^3=P^3Q,$$

we have by selecting the terms in E, E' out of (15) and (16),

$$\left.\begin{aligned}F_{mm_{\prime}}\div\frac{\tau\tau_{\prime}}{\mathfrak{g}}&=-\tfrac{1}{2}EPQ^3\cos 2\epsilon-E'PQ(1-2Q^2)\cos\epsilon'\\ G_{mm_{\prime}}\div\frac{\tau\tau_{\prime}}{\mathfrak{g}}&=\ \ \tfrac{1}{2}EPQ^3\sin 2\epsilon+E'P^3Q\sin\epsilon'\end{aligned}\right\}\quad . \quad . \quad . \quad . \quad (33)$$

It may be shown in a precisely similar way by selecting terms out of (21) that

$$\frac{\mathfrak{N}_{mm_{\prime}}}{C}\div\frac{\tau\tau_{\prime}}{\mathfrak{g}}=\tfrac{1}{2}EQ^4\sin 2\epsilon+E'P^2Q^2\sin\epsilon'\quad . \quad . \quad . \quad . \quad . \quad . \quad . \quad (34)$$

It is worthy of notice that (33) and (34) would be exactly the same, even if we did not put $E_1=E_2=E$; $E'_1=E'_2=E'$; $\epsilon_1=\epsilon_2=\epsilon$; $\epsilon'_1=\epsilon'_2=\epsilon'$, because these new terms depend entirely on the sidereal semi-diurnal and diurnal tides. The new expressions which ought rigorously to give the heights and lagging of the solar semi-diurnal and diurnal tides would only occur in $\mathfrak{L}_{m_{\prime}^2}$.

In the two following sections the results are collected with respect to the rate of change of obliquity and with respect to the tidal friction.

§ 9. *The rate of change of obliquity due to both sun and moon.*

The suffixes m^2, $m_{\prime}^2$, $mm_{\prime}$ to $\frac{di}{dt}$ will indicate the rate of change of obliquity due to the moon alone, to the sun alone, and to the sun and moon jointly.

Then writing for P and Q their values, cos i and sin i, we have by (19) and (29), or by (30),

$$\left.\begin{aligned}\frac{n\mathfrak{g}}{\tau^2}\frac{di_{m'}}{dt}&=\tfrac{1}{2}\sin i\cos i(1-\tfrac{3}{4}\sin^2 i)E\sin 2\epsilon+\tfrac{3}{4}\sin^3 i\cos iE'\sin\epsilon'-\tfrac{3}{8}\sin^3 iE''\sin 2\epsilon''\\ \frac{n\mathfrak{g}}{\tau_{,}^2}\frac{di_{m_{,}'}}{dt}&=\tfrac{1}{2}\sin i\cos i(1-\tfrac{3}{4}\sin^2 i)E\sin 2\epsilon+\tfrac{3}{4}\sin^3 i\cos iE'\sin\epsilon'-\tfrac{3}{8}\sin^3 iE'''\sin 2\epsilon'''\end{aligned}\right\}\quad(35)$$

and by (33) and analogy with (19) and (29)

$$\frac{n\mathfrak{g}}{\tau\tau_{,}}\frac{di_{mm_{,}}}{dt}=-\tfrac{1}{2}\sin^3 i\cos iE\sin 2\epsilon-\sin i\cos^3 iE'\sin\epsilon' \quad . \quad . \quad . \quad . \quad (36)$$

The sum of these three values of $\frac{di}{dt}$ gives the total rate of change of obliquity due both to sun and moon, on the assumption that the three semi-diurnal terms may be grouped together, as also the three diurnal ones.

It will be observed that the joint effect tends to counteract the separate effects; this arises from the fact that, as far as regards the joint effect, the two disturbing bodies may be replaced by rings of matter concentric with the earth but oblique to the equator, and such a ring of matter would cause the obliquity to diminish, as was shown in the abstract of this paper (Proc. Roy. Soc., No. 191, 1878), by general considerations, must be the case.

§ 10. *The rate of tidal friction due to both sun and moon.*

The equation which gives the rate of retardation of the earth's rotation is by (26) $\frac{d\omega_3}{dt}=\frac{\mathfrak{N}}{\mathrm{C}}$; it will however be more convenient henceforward to replace ω_3 by $-n$ and to regard n as a variable, and to indicate by n_0 the value of n at the epoch from which the time is measured.

Generally the suffix 0 to any symbol will indicate its value at the epoch.

Then the equation of tidal friction may be written

$$-\frac{d}{dt}\left(\frac{n}{n_0}\right)=\frac{\mathfrak{N}_{m'}}{\mathrm{C}n_0}+\frac{\mathfrak{N}_{m_{,}'}}{\mathrm{C}n_0}+\frac{\mathfrak{N}_{mm_{,}}}{\mathrm{C}n_0} \quad . \quad . \quad . \quad . \quad . \quad . \quad . \quad . \quad . \quad (37)$$

Then by (22) and (34), in which the semi-diurnal and diurnal terms are grouped together, we have

$$\left.\begin{aligned}\left(\frac{\mathfrak{g}n_0}{\tau^2}\right)\frac{\mathfrak{N}_{m'}}{\mathrm{C}n_0}&=(\cos^2 i+\tfrac{3}{8}\sin^4 i)E\sin 2\epsilon+\sin^2 i(1-\tfrac{3}{4}\sin^2 i)E'\sin\epsilon'=\left(\frac{\mathfrak{g}n_0}{\tau_{,}^2}\right)\frac{\mathfrak{N}_{m_{,}'}}{\mathrm{C}n_0}\\ \left(\frac{\mathfrak{g}n_0}{\tau\tau_{,}}\right)\frac{\mathfrak{N}_{mm_{,}}}{\mathrm{C}n_0}&=\tfrac{1}{2}\sin^4 iE\sin 2\epsilon+\sin^2 i\cos^2 iE'\sin\epsilon'\end{aligned}\right\} \quad . \quad (38)$$

§ 11. *The rate of change of obliquity when the earth is viscous.*

In order to understand the physical meaning of the equations giving the rate of change of obliquity (viz.: (35) and (36) if there be two disturbing bodies, or (29) if there be only one) it is necessary to use numbers. The subject will be illustrated in two cases: first, for the sun, moon, and earth with their present configurations; and secondly, for the case of a planet perturbed by a single satellite. For the first illustration I accordingly take the following data: $g=32\cdot19$ (feet, seconds), the earth's mean radius $a=20\cdot9\times10^6$ feet, the sidereal day ·9973 m. s. days, the sidereal year $=365\cdot256$ m. s. days, the moon's sidereal period 27·3217 m. s. days, the ratio of the earth's mass to that of the moon $\nu=82$, and the unit of time the tropical year 365·242 m. s. days.

Then we have

$$n_0=2\pi\div\cdot9973 \text{ in radians per m. s. day}$$

$$\mathfrak{g}=\frac{2g}{5a}$$

$$\tau=\tfrac{3}{2}\times\tfrac{1}{83}\text{ of }4\pi^2\div(\text{month})^2$$

$$\tau_{,}=\tfrac{3}{2}\text{ of }4\pi^2\div(\text{sidereal year})^2.$$

Then it will be found that

$$\left.\begin{aligned}\frac{\tau^2}{\mathfrak{g}n_0}&=\cdot6598\text{ degrees per million tropical years}\\ \frac{\tau_{,}^2}{\mathfrak{g}n_0}&=\cdot1423\quad\text{,,}\quad\text{,,}\quad\text{,,}\\ \frac{\tau\tau_{,}}{\mathfrak{g}n_0}&=\cdot3064\quad\text{,,}\quad\text{,,}\quad\text{,,}\end{aligned}\right\}\quad . \quad . \qquad (39)$$

These three quantities will henceforth be written u^2, $u_{,}^2$, $uu_{,}$.

For the purpose of analysing the physical meaning of the *differential* equations for $\frac{di}{dt}$ and $\frac{d}{dt}\left(\frac{n}{n_0}\right)$, no distinction will be made between $\frac{\tau^2}{\mathfrak{g}n}$ and $\frac{\tau^2}{\mathfrak{g}n_0}$, &c., for it is here only sought to discover the *rates* of changes. But when we come to *integrate* and find the total changes in a given time, regard will have to be paid to the fact that both τ and n are variables.

For the immediate purpose of this section the numerical values of u^2, $u_{,}^2$, $uu_{,}$ given in (39), will be used.

I will now apply the foregoing results to the particular case where the earth is a viscous spheroid.

Let $\rho=\frac{2gaw}{19\upsilon}$, where υ is the coefficient of viscosity.

Then by the theory of bodily tides as developed in my last paper

$$\left.\begin{array}{l} E=\cos 2\epsilon,\ E'=\cos\epsilon',\ E''=\cos 2\epsilon'',\ E'''=\cos 2\epsilon''' \\ \tan 2\epsilon=\frac{2n}{\rho},\ \tan\epsilon'=\frac{n}{\rho},\ \tan 2\epsilon''=\frac{2\Omega}{\rho},\ \tan 2\epsilon'''=\frac{2\Omega_{,}}{\rho} \end{array}\right\} \quad . \quad . \quad . \quad . \quad (40)$$

Rigorously, we should add to these

$$\left.\begin{array}{l} E_1=\cos 2\epsilon_1,\ E_2=\cos 2\epsilon_2,\ E'_1=\cos\epsilon'_1,\ E'_2=\cos\epsilon'_2 \\ \tan 2\epsilon_1=\frac{2(n-\Omega)}{\rho},\ \tan 2\epsilon_2=\frac{2(n+\Omega)}{\rho},\ \tan\epsilon'_1=\frac{n-2\Omega}{\rho},\ \tan\epsilon'_2=\frac{n+2\Omega}{\rho} \end{array}\right\} \quad . \quad (40')$$

But for the present we classify the three semi-diurnal tides together, as also the three diurnal ones.

Then we have

$$\begin{aligned}\frac{di}{dt}=&\left[\tfrac{1}{4}\sin i\cos i(1-\tfrac{3}{4}\sin^2 i)\sin 4\epsilon+\tfrac{3}{8}\sin^3 i\cos i\sin 2\epsilon'\right](u^2+u_{,}^2)-\tfrac{3}{16}\sin^3 i\sin 4\epsilon'' u^2\\ &-\tfrac{3}{16}\sin^3 i\sin 4\epsilon''' u_{,}^2-(\tfrac{1}{4}\sin^3 i\cos i\sin 4\epsilon+\tfrac{1}{2}\sin i\cos^3 i\sin 2\epsilon')uu_{,}.\end{aligned}$$

Now

$$\begin{aligned}&\tfrac{1}{4}\sin i\cos i(1-\tfrac{3}{4}\sin^2 i)=\tfrac{1}{64}\sin 2i(5+3\cos 2i)=\tfrac{1}{64}(5\sin 2i+\tfrac{3}{2}\sin 4i)\\ &\tfrac{3}{8}\sin^3 i\cos i=\tfrac{3}{32}\sin 2i(1-\cos 2i)=\tfrac{3}{64}(2\sin 2i-\sin 4i)\\ &\tfrac{3}{16}\sin^3 i=\tfrac{3}{64}(3\sin i-\sin 3i),\ \tfrac{1}{4}\sin^3 i\cos i=\tfrac{1}{64}(2\sin 2i-\sin 4i)\\ &\tfrac{1}{2}\sin i\cos^3 i=\tfrac{1}{8}\sin 2i(1+\cos 2i)=\tfrac{4}{64}(2\sin 2i+\sin 4i).\end{aligned}$$

If these transformations be introduced, the equation for $\frac{di}{dt}$ may be written

$$\left.\begin{aligned}64\frac{di}{dt}=&-9(u^2\sin 4\epsilon''+u_{,}^2\sin 4\epsilon''')\sin i+3(u^2\sin 4\epsilon''+u_{,}^2\sin 4\epsilon''')\sin 3i\\ &+\left[(5\sin 4\epsilon+6\sin 2\epsilon')(u^2+u_{,}^2)-(4\sin 4\epsilon+8\sin 2\epsilon')uu_{,}\right]\sin 2i\\ &+\left[(\tfrac{3}{2}\sin 4\epsilon-3\sin 2\epsilon')(u^2+u_{,}^2)+(2\sin 4\epsilon-4\sin 2\epsilon')uu_{,}\right]\sin 4i\end{aligned}\right\} \quad . \quad . \quad (41)$$

Then substituting for u and $u_{,}$ their numerical values (39), and omitting the term depending on the semi-annual tide as unimportant, I find

$$\left.\begin{aligned}64\frac{di}{dt}=&-5\cdot 9378\sin 4\epsilon''\sin i+1\cdot 9793\sin 4\epsilon''\sin 3i\\ &+\{2\cdot 7846\sin 4\epsilon+2\cdot 3611\sin 2\epsilon'\}\sin 2i\\ &+\{1\cdot 8159\sin 4\epsilon-3\cdot 6317\sin 2\epsilon'\}\sin 4i\end{aligned}\right\} \quad . \quad . \quad . \quad (42)$$

The numbers are such that $\frac{di}{dt}$ is expressed in degrees per million years.

The various values which $\frac{di}{dt}$ is capable of assuming as the viscosity and obliquity vary is best shown graphically. In Plate 36, figs. 2 and 3, each curve corresponds to a given degree of viscosity, that is to say to a given value of ϵ, and the ordinates give the values of $\frac{di}{dt}$ as the obliquity increases from 0° to 90°. The scale at the side of each figure is a scale of degrees per hundred million years—*e.g.*, if we had $\epsilon=30°$ and i about 57°, the obliquity would be increasing at the rate of about 3° 45′ per hundred million years.

The behaviour of this family of curves is so very peculiar for high degrees of viscosity, that I have given a special figure (viz.: Plate 36, fig. 3) for the viscosities for which $\epsilon=40°$, 41°, 42°, 43°, 44°.

The peculiarly rapid variation of the forms of the curves for these values of ϵ is due to the rising of the fortnightly tide into prominence for high degrees of viscosity. The matter of the spheroid is in fact so stiff that there is not time in 12 hours or a day to raise more than a very small tide, whilst in a fortnight a considerable lagging tide is raised.

For $\epsilon=44°$ the fortnightly tide has risen to give its maximum effect (*i.e.*, $\sin 4\epsilon''=1$), whilst the effects of the other tides only remain evident in the hump in the middle of the curve. Between $\epsilon=44°$ and 45° the ordinates of the curve diminish rapidly and the hump is smoothed down, so that when $\epsilon=45°$ the curve is reduced to the horizontal axis.

By the theory of the preceding paper,* the values of ϵ when divided by 15 give the corresponding retardation of the bodily semi-diurnal tide—*e.g.*, when $\epsilon=30°$ the tide is two hours late. Also the height of the tide is $\cos 2\epsilon$ of the height of the equilibrium tide of a perfectly fluid spheroid—*e.g.*, when $\epsilon=30°$ the height of tide is reduced by one-half. In the tables given in Part I., Section 7, of the preceding paper, will be found approximate values of the viscosity corresponding to each value of ϵ.

The numerical work necessary to draw these figures was done by means of CRELLE's multiplication table, and as to fig. 2 in duplicate mechanically with a sector; the ordinates were thus only determined with sufficient accuracy to draw a fairly good figure. For the two figures I found 108 values of each of the seven terms of $\frac{di}{dt}$ (nine values of i and twelve of ϵ), and from the seven tables thus formed, the values corresponding to each ordinate of each member of the family were selected and added together.

From this figure several remarkable propositions may be deduced. When the ordinates are positive, it shows that the obliquity tends to increase, and when negative to diminish. Whenever, then, any curve cuts the horizontal axis there is a position of dynamical equilibrium; but when the curve passes from above to below, it

* "On the Bodily Tides of Viscous and Semi-elastic Spheroids," &c., Phil. Trans., 1879, Part I.

is one of stability, and when from below to above, of instability. It follows from this that the positions of stability and instability must occur alternately. When $\epsilon=0$ or 45° (fluidity or rigidity) the curve reduces to the horizontal axis, and every position of the earth's axis is one of neutral equilibrium.

But in every other case the position of 90° of obliquity is not a position of equilibrium, but the obliquity tends to diminish. On the other hand, from $\epsilon=0°$ to about 30° (infinitely small viscosity to tide retardation of two hours), the position of zero obliquity is one of dynamical instability, whilst from then onwards to rigidity it becomes a position of stability.

For viscosities ranging from $\epsilon=0°$ to about $42\frac{1}{4}°$ there is a position of stability which lies between about 50° to 87° of obliquity; and the obliquity of dynamical stability diminishes as the viscosity increases.

For viscosities ranging from $\epsilon=30°$ nearly to about $42\frac{1}{4}°$, there is a second position of dynamical equilibrium, at an obliquity which increases from 0° to about 50°, as the viscosity increases from its lower to its higher value. But this position is one of instability.

From $\epsilon=$ about $42\frac{1}{4}°$ there is only one position of equilibrium, and that stable, viz.: when the obliquity is zero.

If the obliquity be supposed to increase past 90°, it is equivalent to supposing the earth's diurnal rotation reversed, whilst the orbital motion of the earth and moon remains the same as before; but it did not seem worth while to prolong the figure, as it would have no applicability to the planets of the solar system. And, indeed, the figure for all the larger obliquities would hardly be applicable, because any planet whose obliquity increased very much, must gradually make the plane of the orbit of its satellite become inclined to that of its own orbit, and thus the hypothesis that the satellite's orbit remains coincident with the ecliptic would be very inexact.

It follows from an inspection of the figure that for all obliquities there are two degrees of viscosity, one of which will make the rate of change of obliquity a maximum and the other minimum. A graphical construction showed that for obliquities of about 5° to 20°, the degree of viscosity for a maximum corresponds to about $\epsilon=17\frac{1}{2}°$*, whilst that for a minimum to about $\epsilon=40°$. In order, however, to check this conclusion, I determined the values of ϵ analytically when $i=15°$, and when the fortnightly tide (which has very little effect for small obliquities) is neglected. I find that the values are given by the roots of the equation

$$x^3+10x^2+13{\cdot}660x-20{\cdot}412=0, \text{ where } x=3\cos 4\epsilon.$$

This equation has three real roots, of which one gives a hyperbolic cosine, and the

* I may here mention that I found when $\epsilon=17\frac{1}{2}°$, that it would take about a thousand million years for the obliquity to increase from 5° to $23\frac{1}{2}°$, if regard was only paid to this equation of change of obliquity. The equations of tidal friction and tidal reaction will, however, entirely modify the aspects of the case.

other two give $\epsilon=18° \, 15'$ and $\epsilon=41° \, 37'$. This result therefore confirms the geometrical construction fairly well.

It is proper to mention that the expressions of dynamical stability and instability are only used in a modified sense, for it will be seen when the effects of tidal friction come to be included, that these positions are continually shifting, so that they may be rather described as positions of instantaneous stability and instability.

* I will now illustrate the case where there is only one satellite to the planet, and in order to change the point of view, I will suppose that the periodic time of the satellite is so short that we cannot classify the semi-diurnal and diurnal terms together, but must keep them all separate.

Suppose that $n=5\Omega$; then the speeds of the seven tides are proportional to the following numbers, 8, 10, 12 (semi-diurnal); 3, 5, 7 (diurnal); 2 (fortnightly).

These are all the data which are necessary to draw a family of curves similar to those in Plate 36, figs. 2 and 3, because the scale, to which the figure is drawn, is determined by the mass of the satellite, the mass and density of the planet, and the actual velocity of rotation of the planet.

Then by (16) and (29) we have

$$\frac{di}{dt}=\frac{\tau^2}{\mathfrak{g}n}\Big[\tfrac{1}{2}p^7q\sin 4\epsilon_1-p^3q^3(p^2-q^2)\sin 4\epsilon-\tfrac{1}{2}pq^7\sin 4\epsilon_2-\tfrac{3}{2}p^3q^3\sin 4\epsilon''$$
$$+\tfrac{1}{2}p^5q(p^2+3q^2)\sin 2\epsilon'_1-\tfrac{1}{2}pq(p^2-q^2)^3\sin 2\epsilon'-\tfrac{1}{2}pq^5(3p^2+q^2)\sin 2\epsilon'_2\Big]$$

where $p=\cos\frac{i}{2}$ and $q=\sin\frac{i}{2}$

This equation may be easily reduced to the form

$$\frac{di}{dt}=\frac{\tau^2}{\mathfrak{g}n}\frac{1}{128}\sin i\Big\{\quad[10\sin 4\epsilon_1-10\sin 4\epsilon_2+16\sin 2\epsilon'_1-16\sin 2\epsilon'_2-12\sin 4\epsilon'']$$
$$+\cos i[15\sin 4\epsilon_1-4\sin 4\epsilon+15\sin 4\epsilon_2+18\sin 2\epsilon'_1-24\sin 2\epsilon'+18\sin 2\epsilon'_2]$$
$$+\cos 2i[6\sin 4\epsilon_1-6\sin 4\epsilon_2+12\sin 4\epsilon'']$$
$$+\cos 3i[\sin 4\epsilon_1+4\sin 4\epsilon+\sin 4\epsilon_2-2\sin 2\epsilon'_1-8\sin 2\epsilon'-2\sin 2\epsilon'_2]\Big\}$$

which is convenient for the computation of the ordinates of the family of curves which illustrate the various values of $\frac{di}{dt}$ for various obliquities and viscosities.

In Plate 36, fig. 4, the lag (ϵ) of the sidereal semi-diurnal tide is taken as the standard of viscosity. The abscissæ represent the various obliquities of the planet's equator to the plane of the satellite's orbit; the ordinates represent the values of $\frac{di}{dt}$ $\left(\text{the actual scale depending on the value of }\frac{\tau^2}{\mathfrak{g}n}\right)$; and each curve represents one degree of viscosity, viz.: when $\epsilon=10°$, $20°$, 30, $40°$ and $44°$.

* From here to the end of the section was added July 8, 1879.

The computation of the ordinates was done by CRELLE's three-figure multiplication table, and thus the figure does not profess to be very rigorously exact.

This family of curves differs much from the preceding one. For moderate obliquities there is no degree of viscosity which tends to make the obliquity diminish, and thus there is no position of dynamically unstable equilibrium of the system except that of zero obliquity. Thus we see that the decrease of obliquity for small obliquities and large viscosities in the previous case was due to the attraction of the sun on the lunar tides and the moon on the solar tides.

In the present case the position of zero obliquity is never stable, as it was before. The dynamically stable position at a large obliquity still remains as before, but in consequence of the largeness of the ratio $\Omega \div n$ ($\frac{1}{5}$th instead of $\frac{1}{27}$th), this obliquity of dynamical stability is not nearly so great as in the previous case. As the ratio $\Omega \div n$ increases, the position of dynamical stability is one of smaller and smaller obliquity, until when $\Omega \div n$ is equal to a half, zero obliquity becomes stable,—as we shall see later on.

§ 12. *Rate of tidal friction when the earth is viscous.*

If in the same way the equations (37) and (38) be applied to the case where the earth is purely viscous, when the semi-diurnal and diurnal tides are grouped together, we have

$$\left.\begin{aligned} -\frac{d}{dt}\left(\frac{n}{n_0}\right) = (u^2+u_,^2)\left[\tfrac{1}{2}(\cos^2 i+\tfrac{3}{8}\sin^4 i)\sin 4\epsilon+\tfrac{1}{2}\sin^2 i(1-\tfrac{3}{4}\sin^2 i)\sin 2\epsilon'\right] \\ +uu_,\left[\tfrac{1}{4}\sin^4 i\sin 4\epsilon+\tfrac{1}{2}\sin^2 i\cos^2 i\sin 2\epsilon'\right] \end{aligned}\right\} \quad (43)$$

Plate 36, fig. 5, exhibits the various values of $\frac{d}{dt}\left(\frac{n}{n_0}\right)$ for the various obliquities and degrees of viscosity, just as the previous figures exhibited $\frac{di}{dt}$. The calculations were done in the same way as before, after the various functions of the obliquity were expressed in terms of $\cos 2i$ and $\cos 4i$.

The only remarkable point in these curves is that, for the higher degrees of viscosity, the tidal friction rises to a maximum for about 45° of obliquity. The tidal friction rises to its greatest value when $\epsilon=22\frac{1}{2}°$ nearly; this is explained by the fact that by far the largest part of the friction arises from the semi-diurnal tide, which has its greatest effect when $\sin 4\epsilon$ is unity.

§ 13. *Tidal friction and apparent secular acceleration of the moon.*

I now set aside again the hypothesis that the earth is purely viscous, and return to that of there being any kind of lagging tides.

I shall first find at what rate the earth is being retarded when it is moving with its present diurnal rotation, and when the moon is moving in her present orbit, and no distinction will be made between n and n_0; all the secular changes will be considered later.

The numerical data of Section 11 are here used, and the obliquity of the ecliptic $i=23^\circ 28'$; then u and $u_,$ being expressed in radians per tropical year, I find

$$\left.\begin{aligned}\frac{\mathfrak{N}}{C}&=\frac{2\cdot7563}{10^5}E\sin 2\epsilon+\frac{\cdot6143}{10^5}E'\sin\epsilon'\\ \frac{\mathfrak{N}}{Cn}&=\frac{1\cdot1978}{10^8}E\sin 2\epsilon+\frac{\cdot2669}{10^8}E'\sin\epsilon'\end{aligned}\right\}\quad . \quad . \qquad (44)$$

Then integrating the equation (37) and putting $n=n_0$, when $t=0$

$$n=n_0-\frac{\mathfrak{N}}{C}t=n_0\left(1-\frac{\mathfrak{N}}{Cn_0}t\right) \quad . \quad . \quad . \quad . \quad . \quad . \quad . \qquad (45)$$

Integrating a second time, we find that a fixed meridian in the earth has fallen behind the place it would have had, if the rotation had not been retarded, by $\frac{1}{2}\frac{\mathfrak{N}t^2}{C}\cdot\frac{648000}{\pi}$ seconds of arc. And at the end of a century it is behind time $1900\cdot27E\sin 2\epsilon+423\cdot49E'\sin\epsilon'$ m. s. seconds of time.

If the earth were purely viscous, and when $\epsilon=17^\circ\ 30'$* (which by Section 11 causes the rate of change of obliquity to be a maximum), I find that at the end of a century the earth is behind time in its rotation by 17 minutes 5 seconds.

By substitution from the second of (44), equation (45) may be written in the form

$$n=n_0\left(1-\frac{1\cdot1978}{10^8}tE\sin 2\epsilon-\frac{\cdot2669}{10^8}tE'\sin\epsilon'\right). \quad . \quad . \quad . \quad . \qquad (46)$$

which in the supposed case of pure viscosity when $\epsilon=17^\circ\ 30'$ becomes

$$n=n_0\left(1-\frac{\cdot006460}{10^6}t\right) \quad . \quad . \quad . \quad . \quad . \quad . \quad . \quad . \qquad (47)$$

All these results would, however, cease to be even approximately true after a few millions of years.

The effect of the failure of the earth to keep true time is to cause an apparent acceleration of the moon's motion; and if the moon's motion were really unaffected by

* This calculation was done before I perceived that I had not chosen that degree of viscosity which makes the tidal friction a maximum, but as all the other numerical calculations have been worked out for this degree of viscosity I adhere to it here also.

the tides in the earth, there would be an apparent acceleration of the moon in a century of

$$1043''\cdot 28 E \sin 2\epsilon + 232''\cdot 50 E' \sin \epsilon' \quad . \quad . \quad . \quad . \quad (48)$$

for the moon moves over $0''\cdot 5490$ of her orbit in one second of time.

This apparent acceleration would however be considerably diminished by the effects of tidal reaction on the moon, which will now be considered.

§ 14. *Tidal reaction on the moon.**

The action of the tides on the moon gives rise to a small force tangential to the orbit accelerating her linear motion. The spiral described by the moon about the earth will differ insensibly from a circle, and therefore we may assume throughout that the centrifugal force of the earth's and moon's orbital motion round their common centre of inertia is equal and opposite to the attraction between them.

We shall now find the tangential force on the moon in terms of the couples which we have already found acting on the earth. Those couples consist of the sum of three parts, viz.: that due (i) to the moon alone, (ii) to the sun alone, and (iii) to the action of the sun on the lunar tides and of the moon on the solar tides, the latter two being equal *inter se*.

Now since action and reaction are equal and opposite, therefore the only parts of these couples which correspond with the tangential force on the moon are those which arise from (i), and one-half those which arise from (iii).

We may thus leave the sun out of account if we suppose the earth only to be acted on by the couples $\mathfrak{L}_{m} + \frac{1}{2}\mathfrak{L}_{mm_1}$, $\mathfrak{M}_{m} + \frac{1}{2}\mathfrak{M}_{mm_1}$, $\mathfrak{N}_{m} + \frac{1}{2}\mathfrak{N}_{mm_1}$; these couples will be called $\mathfrak{L}'$, $\mathfrak{M}'$, $\mathfrak{N}'$, and the part of the change of obliquity which is due to $\mathfrak{L}'$, $\mathfrak{M}'$ will be called $\frac{di'}{dt}$.

Let r and $-\Omega$ be the moon's distance, and angular velocity at any time, and ν the ratio of the earth's mass to the moon's.

Let T be the force which acts on the moon perpendicular to her radius vector, in the direction of her motion.

From the equality of action and reaction, it follows that Tr must be equal to the couple which is produced by the moon's action on the tides in the earth, acting in the direction tending to retard the earth's diurnal rotation about the normal to the ecliptic. Referring to Plate 36, fig. 1, we see that the direction cosines of this normal are $-\sin i \cos n$, $-\sin i \sin n$, $\cos i$; hence

$$\text{Tr} = -\sin i(\mathfrak{L}' \cos n + \mathfrak{M}' \sin n) + \mathfrak{N}' \cos i.$$

* This section has been partly rewritten and rearranged since the paper was presented. (Dec. 19, 1878.)

But by (17) and (18)

$$\frac{\mathfrak{L}'}{C} = \quad (F_{m^2}+\tfrac{1}{2}F_{mm_,}) \sin n + (G_{m^2}+\tfrac{1}{2}G_{mm_,}) \cos n$$

$$\frac{\mathfrak{M}'}{C} = -(F_{m^2}+\tfrac{1}{2}F_{mm_,}) \cos n + (G_{m^2}+\tfrac{1}{2}G_{mm_,}) \sin n.$$

Hence

$$\frac{\mathfrak{L}'}{C} \cos n + \frac{\mathfrak{M}'}{C} \sin n = G_{m^2}+\tfrac{1}{2}G_{mm_,} = -n\frac{di'}{dt}.$$

Thus

$$\mathrm{Tr} = C\left\{\frac{\mathfrak{N}'}{C}\cos i + n \sin i \frac{di'}{dt}\right\} \quad . \quad . \quad . \qquad (49)$$

In order to apply the ordinary formula for the motion of the moon, the earth must be reduced to rest, and therefore T must be augmented by the factor $(M+m)\div M$. Then if ϑ be the moon's longitude, the equation of motion of the moon is

$$m\frac{d}{dt}\left(\mathrm{r}^2\frac{d\vartheta}{dt}\right) = \frac{M+m}{M}\mathrm{Tr} \quad . \quad . \quad . \quad . \quad . \quad . \quad . \quad . \quad . \quad . \qquad (50)$$

But since the orbit is approximately circular $\dfrac{d\vartheta}{dt} = \Omega$.

Also $m = C \div \tfrac{2}{5}\nu a^2$, and $\dfrac{M+m}{M} = \dfrac{1+\nu}{\nu}$.

Therefore by (49) and (50)

$$\frac{d(\Omega \mathrm{r}^2)}{dt} = \tfrac{2}{5}\nu a^2 \frac{1+\nu}{\nu}\left\{\frac{\mathfrak{N}'}{C}\cos i + n \sin i \frac{di'}{dt}\right\}$$

Now let $\xi = \left(\dfrac{\Omega_0}{\Omega}\right)^{\frac{1}{3}}$, whence $\Omega^2 = \Omega_0{}^2 \div \xi^6$.

The suffix 0 to Ω indicates the value of Ω when the time is zero, and no confusion will arise by this second use of the symbol ξ.

But since the centrifugal force is equal to the attraction between the two bodies, and the orbit is circular, therefore $\Omega^2\mathrm{r}^3 = M+m$.

So that $\Omega_0{}^2\mathrm{r}^3 = (M+m)\xi^6$.

Therefore

$$\mathrm{r}^2 = (M+m)^{\frac{2}{3}}\xi^4\Omega_0{}^{-\frac{4}{3}}, \text{ and } \Omega\mathrm{r}^2 = (M+m)^{\frac{2}{3}}\Omega_0{}^{-\frac{1}{3}}\xi$$

and hence

$$\frac{d}{dt}(\Omega\mathrm{r}^2) = (M+m)^{\frac{2}{3}}\Omega_0{}^{-\frac{1}{3}}\frac{d\xi}{dt}$$

But $M+m = ga^2\dfrac{1+\nu}{\nu}$, because M and m are here measured in astronomical units of mass.

Therefore our equation may be written

$$\left(ga^2\frac{1+\nu}{\nu}\right)^{\frac{1}{3}}\Omega_0^{-\frac{1}{3}}\frac{d\xi}{dt}=\tfrac{2}{5}a^2(1+\nu)\left\{\frac{\mathfrak{N}}{\mathrm{C}}\cos i+n\sin i\frac{di'}{dt}\right\}$$

Now let

$$s=\tfrac{2}{5}\left[\left(\frac{a}{g}\right)^2\nu^2(1+\nu)\right]^{\frac{1}{3}},\text{ and let } sn_0\Omega_0^{\frac{1}{3}}=\frac{1}{\mu},\text{ and let } N=\frac{n}{n_0}\quad . \quad . \quad . \qquad (51)$$

And we have

$$\mu\frac{d\xi}{dt}=\frac{\mathfrak{N}'}{\mathrm{C}n_0}\cos i+N\sin i\frac{di'}{dt} \qquad . \quad (52)$$

It is not hard to show that the moment of momentum of the orbital motion of the two bodies is $\mathrm{C}\div s\Omega^{\frac{1}{3}}$, and that of the earth's rotation is obviously $\mathrm{C}n$. Hence $sn\Omega^{\frac{1}{3}}$ is the ratio of the two momenta, and μ is the ratio of the two momenta at the fixed moment of time, which is the epoch.

In the similar equation expressive of the rate of change in the earth's orbital motion round the sun, it is obvious that the orbital moment of momentum is so very large compared with the earth's moment of momentum of rotation, that μ is very large and the earth's mean distance from the sun remains sensibly constant (see Section 19).

Then by (16) and (29), remembering that

$$p=\cos\frac{i}{2},\ q=\sin\frac{i}{2},\ \frac{di_{m^2}}{dt}=-\frac{\mathfrak{G}_{m^2}}{n},\text{ and } N=\frac{n}{n_0},$$

we have

$$\begin{aligned}N\sin i\frac{di_{m^2}}{dt}=\frac{\tau^2}{\mathfrak{g}n_0}2pq[&E_1p^7q\sin 2\epsilon_1-E2p^3q^3(p^2-q^2)\sin 2\epsilon-E_2pq^7\sin 2\epsilon_2\\&+E'_1p^5q(p^2+3q^2)\sin\epsilon'_1-E'pq(p^2-q^2)^3\sin\epsilon'-E'_2pq^5(3p^2+q^2)\sin\epsilon'_2\\&-E''3p^3q^3\sin 2\epsilon'']\quad . \quad . \quad . \quad . \quad . \quad . \quad . \quad . \quad . \quad . \quad . \quad . \quad . \quad . \quad . \qquad (53)\end{aligned}$$

And by (21)

$$\begin{aligned}\cos i\frac{\mathfrak{N}_{m^2}}{\mathrm{C}n_0}=\frac{\tau^2}{\mathfrak{g}n_0}(p^2-q^2)[&E_1p^8\sin 2\epsilon_1+E4p^4q^4\sin 2\epsilon+E_2q^8\sin 2\epsilon_2\\&+E'_12p^6q^2\sin\epsilon'_1+E'2p^2q^2(p^2-q^2)^2\sin\epsilon'+E'_22p^2q^6\sin\epsilon'_2]\ . \qquad (54)\end{aligned}$$

By (33) and (34), and remembering to take the halves of $\mathfrak{G}_{mm_{\prime}}$ and $\mathfrak{N}_{mm_{\prime}}$, and that $\sin i=Q$, $\cos i=P$

$$N\sin i\left(\tfrac{1}{2}\frac{di_{mm_{\prime}}}{dt}\right)=-\frac{\tau\tau_{\prime}}{\mathfrak{g}n_0}Q[\tfrac{1}{4}EPQ^3\sin 2\epsilon+\tfrac{1}{2}E'P^3Q\sin\epsilon']\quad . \quad . \quad . \quad . \qquad (55)$$

$$\cos i\tfrac{1}{2}\frac{\mathfrak{N}_{mm_{\prime}}}{\mathrm{C}n_0}=\frac{\tau\tau_{\prime}}{\mathfrak{g}n_0}P[\tfrac{1}{4}EQ^4\sin 2\epsilon+\tfrac{1}{2}E'P^2Q^2\sin\epsilon']\ . \quad . \quad . \quad . \quad . \qquad (56)$$

Now to obtain $\mu\frac{d\xi}{dt}$, we have to add the last four expressions together, and we observe that the last two cut one another out, so that the expression for $\frac{d\xi}{dt}$ is independent of the solar tides; also the terms in $\sin 2\epsilon$, $\sin\epsilon'$ cut one another out in the sum of the first two expressions, and hence it follows that $\frac{d\xi}{dt}$ is independent of the sidereal semi-diurnal and diurnal terms.

Thus we have

$$\mu\frac{d\xi}{dt}=\frac{\tau^2}{\mathfrak{g}n_0}[E_1p^8\sin 2\epsilon_1-E_2q^8\sin 2\epsilon_2+4E'_1p^6q^2\sin\epsilon'_1-4E'_2p^2q^6\sin\epsilon'_2-6E''p^4q^4\sin 2\epsilon''] \quad (57)$$

This equation will be referred to hereafter as that of tidal reaction.* From its form we see that the tides of speeds $2(n+\Omega)$, $n+2\Omega$, and 2Ω tend to make the moon approach the earth, whilst the other tides tend to make it recede.

Then if, as in previous cases, we put $E_1=E_2=E$; $E'_1=E'_2=E'$; $\epsilon_1=\epsilon_2=\epsilon$; $\epsilon'_1=\epsilon'_2=\epsilon'$ (which is justifiable so long as the moon's orbital motion is slow compared with that of the earth's rotation), we have, after noticing that

$$p^8-q^8=(p^2-q^2)(p^4+q^4)=\cos i(1-\tfrac{1}{2}\sin^2 i)$$
$$4p^6q^2-4p^2q^6=4p^2q^2(p^2-q^2)=\sin^2 i\cos i$$
$$6p^4q^4=\tfrac{3}{8}\sin^4 i$$

$$\mu\frac{d\xi}{dt}=\frac{\tau^2}{\mathfrak{g}n_0}[\cos i(1-\tfrac{1}{2}\sin^2 i)E\sin 2\epsilon+\sin^2 i\cos iE'\sin\epsilon'-\tfrac{3}{8}\sin^4 iE''\sin 2\epsilon''] \quad . \quad (58)$$

Now if the present values of n, Ω, i be substituted in this equation (58) (*i.e.*, with the present day, month, and obliquity), and if the tropical year be the unit of time, it will be found that

$$10^{10}\frac{d\xi}{dt}=\frac{1}{\xi^{12}}(24{\cdot}27E\sin 2\epsilon+4{\cdot}18E'\sin\epsilon'-{\cdot}271E''\sin 2\epsilon'')$$

ξ^{12} enters into this equation because τ varies as Ω^2 and therefore as ξ^{-6}.

But we may here put $\xi=1$, because at present we only want the instantaneous rate of increase of Ω.

Now $\frac{d\xi}{dt}=-\tfrac{1}{3}\Omega^{-\frac{4}{3}}\Omega_0^{\frac{1}{3}}\frac{d\Omega}{dt}=-\frac{1}{3\Omega_0}\frac{d\Omega}{dt}$ when $\Omega=\Omega_0$; hence multiplying the equation by $3\Omega_0$ we have at the present time

$$-10^{10}\frac{d\Omega}{dt}=6115E\sin 2\epsilon+1053E'\sin\epsilon'-68{\cdot}28E''\sin 2\epsilon'' \quad . \quad . \quad (59)$$

in radians per annum.

* In a future paper on the perturbations of a satellite revolving about a viscous primary, I shall obtain this equation by the method of the disturbing function.

Then if for the moment we call the right-hand of this equation k, we have $\Omega=\Omega_0-k\frac{t}{10^{10}}$. Integrating a second time, we find that the moon has fallen behind her proper place in her orbit $\frac{1}{2}t^2\frac{k}{10^{10}}\cdot\frac{648000}{\pi}$ seconds of arc in the time t. Put t equal a century, and substitute for k, and it will then be found that the moon lags in a century

$$630{\cdot}7E\sin 2\epsilon+108{\cdot}6E'\sin\epsilon'-7{\cdot}042E''\sin 2\epsilon'' \text{ seconds of arc} \quad . \quad . \quad (60)$$

But it was shown in Section 13 (48) that the moon, if unaffected by tidal reaction, would have been apparently accelerated $1043{\cdot}3E\sin 2\epsilon+232{\cdot}5E'\sin\epsilon'$ seconds of arc in a century.

Hence taking the difference of these two, we find that there is an apparent acceleration of the moon's motion of

$$412{\cdot}6E\sin 2\epsilon+123{\cdot}9E'\sin\epsilon'+7{\cdot}042E''\sin 2\epsilon'' \quad . \quad . \quad . \quad . \quad . \quad . \quad (61)$$

seconds of arc in a century.

Now according to ADAMS and DELAUNAY, there is at the present time an unexplained acceleration of the moon's motion of about 4″ in a century. For the present I will assume that the whole of this 4″ is due to the bodily tidal friction and reaction, leaving nothing to be accounted for by ocean tidal friction and reaction, to which the whole has hitherto been attributed. Then we must have

$$412{\cdot}6E\sin 2\epsilon+123{\cdot}9E'\sin\epsilon'+7{\cdot}042E''\sin 2\epsilon''=4 \quad . \quad . \quad . \quad . \quad . \quad (62)$$

This equation gives a relation which must subsist between the heights E, E', E'', of the semi-diurnal, diurnal, and fortnightly bodily tides, and their retardations ϵ, ϵ', ϵ'', in order that the observed amount of tidal friction may not be exceeded. But no further deduction can be made, without some assumption as to the nature of the matter constituting the earth.

I shall first assume then that the matter is purely viscous, so that $E=\cos 2\epsilon$, $E'=\cos\epsilon'$, $E''=\cos 2\epsilon''$, and $\tan 2\epsilon=\frac{2n}{\rho}$, $\tan\epsilon'=\frac{n}{\rho}$, $\tan 2\epsilon''=\frac{2\Omega}{\rho}$. The equation then becomes

$$412{\cdot}6\sin 4\epsilon+123{\cdot}9\sin 2\epsilon'+7{\cdot}042\sin 4\epsilon''=8 \quad . \quad . \quad . \quad . \quad . \quad . \quad (63)$$

If the values of ϵ, ϵ', ϵ'' be substituted, we get an equation of the sixth degree for ρ, but it will not be necessary to form this equation, because the question may be more simply treated by the following approximation.

There are obviously two solutions of the equation, one of which represents that the earth is very nearly fluid, and the other that it is very nearly rigid.

In the first case, that of approximate fluidity, ϵ, ϵ', ϵ'' are very small, and therefore

$$\sin 4\epsilon = 4\epsilon,\ \sin 2\epsilon' = 2\epsilon' = 2\epsilon,\ \sin 4\epsilon'' = 4\epsilon'' = 4\frac{\Omega}{n}\epsilon = \frac{4}{27\cdot 32}\epsilon$$

Hence

$$(1650 + 248 + \tfrac{4}{27\cdot 32} \text{ of } 7\cdot 04)\epsilon = 8$$

whence

$$\epsilon = \tfrac{1}{237} = 14'$$

That is to say, the semi-diurnal tide only lags by the small angle 14′. But this is not the solution which is interesting in the case of the earth, for we know that the earth does not behave approximately as a fluid body.

In the other solution, 2ϵ and ϵ' approach 90°, so that ρ is small; hence

$$\sin 4\epsilon = \frac{4n\rho}{\rho^2 + 4n^2} = \frac{\rho}{n},\ \sin 2\epsilon' = \frac{2n\rho}{\rho^2 + n^2} = \frac{2\rho}{n} \text{ very nearly, and } \sin 4\epsilon'' = \frac{4\Omega\rho}{\rho^2 + 4\Omega^2}$$

Hence we have

$$412\cdot 6\left(\frac{\rho}{n}\right) + 123\cdot 9\left(\frac{2\rho}{n}\right) + 7\cdot 042\frac{4\Omega\rho}{\rho^2 + 4\Omega^2} = 8$$

Put $\frac{\rho}{2\Omega} = x$, so that $x = \cot 2\epsilon''$; then substituting for $\frac{\Omega}{n}$ its value $\frac{1}{27\cdot 32}$, we have

$$\frac{1320\cdot 7}{27\cdot 32}x + 7\cdot 042\frac{2x}{x^2+1} = 8$$

whence

$$x^3 - \cdot 1655x^2 + 1\cdot 2921x - \cdot 1655 = 0$$

This equation has two imaginary roots, and one real one, viz.: ·12858. Hence the desired solution is given by $\cot 2\epsilon'' = \cdot 12858$; and $2\epsilon'' = \frac{1}{2}\pi - 7° 20'$, and the corresponding values of 2ϵ and ϵ' are $2\epsilon = \frac{1}{2}\pi - 16'$, and $\epsilon' = \frac{1}{2}\pi - 32'$. If these values for ϵ, ϵ', ϵ'' be used in the original equation (63), they will be found to satisfy it very closely; and it appears that there is a true retardation of the moon of 3″·1 in a century, whilst the lengthening of the day would make an apparent acceleration of 7″·1,—the difference of the two being the observed 4″.

With these values the semi-diurnal and diurnal ocean-tides are, according to the equilibrium theory of ocean-tides, sensibly the same as those on a rigid nucleus, whilst the fortnightly tide is reduced to $\sin 2\epsilon''$ or ·992 of its theoretical amount; and the time of high tide is accelerated by $\frac{\pi}{4\Omega} - \frac{\epsilon''}{\Omega}$, or $6\frac{1}{2}$ hours in advance of its theoretical time.*

* In the abstract of this paper (Proc. Roy. Soc., No. 191, 1878) the height and lag of the bodily tide were accidentally given instead of the height and acceleration of the ocean tide.

If these values be substituted in the equation giving the rate of variation of the obliquity, it will be found that the obliquity must be decreasing at the rate of ·00197° per million years, or 1° in 500 million years. Thus in 100 million years it would only decrease by 12′. So, also, it may be shown that the moon's sidereal period is being increased by 2 hours 20 minutes in 100 million years.

Lastly, the earth considered as a clock is losing 13 seconds in a century.

There is another supposition as to the physical constitution of the earth, which will lead to interesting results.

If the earth be elastico-viscous, then for the semi-diurnal and diurnal tides it might behave nearly as though it were perfectly elastic, whilst for the fortnightly tide it might behave nearly as though it were perfectly viscous. With the law of elastico-viscosity used in my previous paper,* it is not possible to satisfy these conditions very exactly. But there is no reason to suppose that that law represents anything but an ideal sort of matter; it is as likely that the degradation of elasticity immediately after straining is not so rapid as that law supposes. I shall therefore take a limiting case, and suppose that, for the semi-diurnal and diurnal tides, the earth is perfectly elastic, whilst for the fortnightly one it is perfectly viscous. This hypothesis, of course, will give results in excess of what is rigorously possible, at least without a discontinuity in the law of degradation of elasticity.

It is accordingly assumed that the semi-diurnal and diurnal bodily tides do not lag, and therefore $\epsilon=\epsilon'=0$; whilst the fortnightly tide does lag, and $E''=\cos 2\epsilon''$.

Thus by (38) there is no tidal friction, and by (60) there is a true acceleration of the moon's motion of $\frac{1}{2}$ of $7{\cdot}042 \sin 4\epsilon''$ seconds of arc in a century. Then if we take the most favourable case, namely, when $\epsilon''=22° \; 30'$, there is a true secular acceleration of $3''{\cdot}521$ per century.

It follows, therefore, that the whole of the observed secular acceleration of the moon might be explained by this hypothesis as to the physical constitution of the earth. On this hypothesis the fortnightly ocean tides should amount to sin 22° 30′, or ·38 of its theoretical height on a rigid nucleus, and the time of high water should be accelerated by 1 day 17 hours. Again, by (35) $\frac{di}{dt}=-\frac{3}{16}u^2\sin^3 i$, from whence it may be shown that the obliquity of the ecliptic would be decreasing at the rate of 1° in 128 million years.

The conclusion to be drawn from all these calculations is that, at the present time, the bodily tides in the earth, except perhaps the fortnightly tide, must be exceedingly small in amount; that it is utterly uncertain how much of the observed 4″ of acceleration of the moon's motion must be referred to the moon itself, and how much to

* Namely, that if the solid be strained, the stress required to maintain it in the strained configuration diminishes in geometrical progression as the time, measured from the epoch of straining, increases in arithmetical progression. See Section 8 of the paper on "Bodily Tides," &c., Phil. Trans., Part I., 1879.

the tidal friction, and accordingly that it is equally uncertain at what rate the day is at present being lengthened; lastly, that if there is at present any change in the obliquity to the ecliptic, it must be very slowly decreasing.

The result of this hypothesis of elastico-viscosity appears to me so curious that I shall proceed to show what might possibly have been the state of things a very long time ago, if the earth had been perfectly elastic for the tides of short period, but viscous for the fortnightly tide.

There will now be no tidal friction, and the length of day remains constant. The equation of tidal reaction reduces to

$$\mu\frac{d\xi}{dt} = -\frac{u^2}{\xi^{12}}\tfrac{3}{16}\sin^4 i\sin 4\epsilon''$$

Here u^2 is a constant, being the value of $\frac{\tau^2}{\mathfrak{g}n_0}$ at the epoch; and $u^2\div\xi^{12}$ is the value of $\frac{\tau^2}{\mathfrak{g}n_0}$ at the time t.

The equation giving the rate of change of obliquity becomes

$$\frac{di}{dt} = -\frac{u^2}{\xi^{12}}\tfrac{3}{16}\sin^3 i\sin 4\epsilon''$$

Dividing the latter by the former, we have*

$$\sin i\,di = \mu d\xi$$

And by integration

$$\cos i = \cos i_0 - \mu(\xi - 1)$$

If we look back long enough in time, we may find $\xi = 1{\cdot}01$, and μ being $4{\cdot}007$, we have

$$\cos i = \cos i_0 - {\cdot}04007$$

Taking $i_0 = 23°\,28'$, we find $i = 28°\,40'$.

This result is independent of the degree of viscosity. When, however, we wish to find how long a period is requisite for this amount of change, some supposition as to viscosity is necessary. The time cannot be less than if $\sin 4\epsilon'' = 1$, or $\epsilon'' = 22°\,30'$, and we may find a rough estimate of the time by writing the equation of tidal reaction

$$\mu\frac{d\xi}{dt} = -\tfrac{3}{16}\frac{u^2}{\xi^{12}}\sin^4 \mathrm{I},$$

where I is constant and equal to 24°, suppose. Then integrating we have

$$\mu(\xi^{13} - 1) = -t\tfrac{39}{16}u^2\sin^4 \mathrm{I},$$

or

$$t = -\tfrac{16}{39}\frac{\mu}{u^2}\,\mathrm{cosec}^4\,\mathrm{I}(\xi^{13} - 1).$$

* Concerning the legitimacy of this change of variable, see the following section.

When $\xi=1\cdot01$, we find from this that $-t=720$ million years, and that the length of the month is 28·15 m. s. days. Hence, if we look back 700 million years or more, we might find the obliquity 28° 40′, and the month 28·15 m. s. days, whilst the length of day might be nearly constant. It must, however, be reiterated, that on account of our assumptions the change of obliquity is greater than would be possible, whilst the time occupied by the change is too short. In any case, any change in this direction approaching this in magnitude seems excessively improbable.

PART II.

§ 15. *Integration of the differential equations for secular changes in the variables in the case of viscosity.*[*]

It is now supposed that the earth is a purely viscous spheroid, and I shall proceed to find the changes which would occur in the obliquity to the ecliptic and the lengths of the day and month when very long periods of time are taken into consideration.

I have been unable to find even an approximate general analytical solution of the problem, and have therefore worked the problem by a laborious arithmetical method, when the earth is supposed to have a particular degree of viscosity.

The viscosity chosen is such that, with the present length of day, the semi-diurnal tide lags by 17° 30′. It was shown above that this viscosity makes the rate of change of obliquity nearly a maximum.[†] It does not follow that the whole series of changes will proceed with maximum velocity, yet this supposition will, I think, give a very good idea of the minimum time, and of the nature of the changes which may have occurred in the course of the development of the moon-earth system.

The three semi-diurnal tides will be supposed to lag by the same amount and to be reduced in the same proportion; as also will be the three diurnal tides.

There are three simultaneous differential equations to be treated, viz.: those giving (1) the rate of change of the obliquity of the ecliptic, (2) the rate of alteration of the earth's diurnal rotation, (3) the rate of tidal reaction on the moon. They will be referred to hereafter as the *equations of obliquity, of friction, and reaction* respectively.

To write these equations more conveniently a partly new notation is advantageous, as follows:—

The suffix 0 to any symbol denotes the initial value of the quantity in question.

Let $u^2=\frac{\tau_0^2}{\mathfrak{g}n_0}$, $u_,^2=\frac{\tau_,^2}{\mathfrak{g}n_0}$, $uu_,=\frac{\tau_0\tau_,}{\mathfrak{g}n_0}$; these three quantities are constant.

* This section has been rearranged, partly rewritten, and recomputed since the paper was presented. The alterations were made on December 19, 1878.

† If I had to make the choice over again I should choose a slightly greater viscosity as being more interesting.

Since the tidal reaction on the sun is neglected, $\tau_{,}$ is a constant, and since τ varies as Ω^2 (and therefore as ξ^{-6}) ; hence

$$\frac{\tau^2}{\mathfrak{g}n}=\frac{n_0}{n}\frac{u^2}{\xi^{12}},\quad \frac{\tau_{,}^2}{\mathfrak{g}n}=\frac{n_0}{n}u_{,}^2,\quad \frac{\tau\tau_{,}}{\mathfrak{g}n}=\frac{n_0}{n}\frac{uu_{,}}{\xi^6}$$

Let ρ be equal to $\frac{2gaw}{19v}$, where v is the coefficient of viscosity of the earth. Then according to the theory developed in my paper on tides*

$$\tan 2\epsilon=\frac{2n}{\rho},\ \tan\epsilon'=\frac{n}{\rho},\ \tan 2\epsilon''=\frac{2\Omega}{\rho}\quad . \quad . \quad . \quad . \quad . \quad . \quad . \quad (64)$$

To simplify the work, terms involving the fourth power of the sine of the obliquity will be neglected.

Now let

$$\left.\begin{aligned} &P=\tfrac{1}{4}\log_{10}e,\ Q=\tfrac{3}{8}\sin^2 i\log_{10}e,\ R=\tfrac{3}{16}\frac{\sin^2 i}{\cos i}\log_{10}e=\tfrac{1}{2}Q\sec i\\ &U=\tfrac{1}{4}\sin^2 i\log_{10}e,\ V=\frac{\tfrac{1}{2}\cos^2 i}{1-\tfrac{3}{4}\sin^2 i}\log_{10}e\\ &W=\tfrac{1}{2}\cos^2 i,\ X=\tfrac{1}{2}\sin^2 i\cos i,\ Z=\tfrac{1}{2}\sin^2 i\cos^2 i \end{aligned}\right\}\quad . \quad . \quad (65)$$

Also let $sn_0\Omega_0^{\frac{1}{3}}=\frac{1}{\mu}$, $\frac{n}{n_0}=N$; and it may be called to mind that $\xi=\left(\frac{\Omega_0}{\Omega}\right)^{\frac{1}{3}}$, $s=\frac{2}{5}\left[\left(\frac{av}{g}\right)^2(1+\nu)\right]^{\frac{1}{3}}$.

The terms depending on the semi-annual tide will be omitted throughout.

With this notation the equation of obliquity (35) and (36) may be written,

$$\begin{aligned}\log_{10}e\frac{di}{dt}=\sin i\cos i(1-\tfrac{3}{4}\sin^2 i)\Big[&\left(\frac{u^2}{\xi^{12}}+u_{,}^2\right)(P\sin 4\epsilon+Q\sin 2\epsilon')\\ &-\frac{uu_{,}}{\xi^6}(U\sin 4\epsilon+V\sin 2\epsilon')-\frac{u^2}{\xi^{12}}R\sin 4\epsilon''\Big]\quad . \quad . \quad . \quad . \quad . \quad . \quad (66)\end{aligned}$$

The equation (43) of friction becomes

$$-\frac{dN}{dt}=\left(\frac{u^2}{\xi^{12}}+u_{,}^2\right)(W\sin 4\epsilon+X\sin 2\epsilon')+\frac{uu_{,}}{\xi^6}Z\sin 2\epsilon'\quad . \quad . \quad . \quad (67)$$

And by (58), Section 14, the equation of reaction becomes

$$\mu\frac{d\xi}{dt}=\frac{u^2}{\xi^{12}}(W\sin 4\epsilon+X\sin 2\epsilon')\quad . \quad . \quad . \quad . \quad . \quad . \quad . \quad . \quad (68)$$

* Phil. Trans., 1879, Part I.

This is the third of the simultaneous differential equations which have to be treated. The four variables involved are i, N, ξ, t, which give the obliquity, the earth's rotation, the square root of the moon's distance and the time. Besides where they are involved explicitly, they enter implicitly in Q, R, U, V, W, X, Z, $\sin 4\epsilon$, $\sin 2\epsilon'$, $\sin 4\epsilon''$.

Q, R, &c., are functions of the obliquity i only, but P is a constant. Also $\sin 4\epsilon = \frac{4n\rho}{4n^2+\rho^2} = \frac{4n_0\rho N}{4n_0^2N^2+\rho^2}$, $\sin 2\epsilon' = \frac{2n_0\rho N}{n_0^2N^2+\rho^2}$, $\sin 4\epsilon'' = \frac{4\Omega_0\rho\xi^3}{4\Omega_0^2+\rho^2\xi^6}$. I made several attempts to solve these equations by retaining the time as independent variable, and substituting for ξ and N approximate values, but they were all unsatisfactory, because of the high powers of ξ which occur, and no security could be felt that after a considerable time the solutions obtained did not differ a good deal from the true one. The results, however, were confirmatory of those given hereafter.

The method finally adopted was to change the independent variable from t to ξ. A new equation was thus formed between N and ξ, which involved the obliquity i only in a subordinate degree, and which admitted of approximate integration. This equation is in fact that of conservation of moment of momentum, modified by the effects of the solar tidal friction. Afterwards the time and the obliquity were found by the method of quadratures. As, however, it was not safe to push this solution beyond a certain point, it was carried as far as seemed safe, and then a new set of equations were formed, in which the final values of the variables, as found from the previous integration, were used as the initial values. A similar operation was carried out a third and fourth time. The operations were thus divided into a series of periods, which will be referred to as periods of integration. As the error in the final values in any one period is carried on to the next period, the error tends to accumulate; on this account the integration in the first and second periods was carried out with greater accuracy than would in general be necessary for a speculative inquiry like the present one. The first step is to form the approximate equation of conservation of moment of momentum above referred to.

Let $A = W\sin 4\epsilon + X\sin 2\epsilon'$, $B = Z\sin 2\epsilon'$.

Then the equations of friction (67) and reaction (68) may be written,

$$-n_0\mathfrak{g}\frac{dN}{dt} = \left(\frac{\tau_0^2}{\xi^{12}} + \tau_,^2\right)A + \frac{\tau_0\tau_,}{\xi^6}B \quad . \quad . \quad . \quad . \quad . \quad . \quad . \quad (69)$$

$$n_0\mathfrak{g}\mu\frac{d\xi}{dt} = \frac{\tau_0^2}{\xi^{12}}A \quad . \quad . \quad . \quad . \quad . \quad . \quad . \quad . \quad . \quad . \quad (70)$$

We now have to consider the proposed change of variable from t to ξ.

The full expression for $\frac{dN}{dt}$ contains a number of periodic terms; $\frac{d\xi}{dt}$ also contains terms which are co-periodic with those in $\frac{dN}{dt}$. Now the object which is here in view is to determine the increase in the average value of N per unit increase of the average value of ξ. The proposed new independent variable is therefore not ξ, but it is the average value of ξ; but as no occasion will arise for the use of ξ as involving periodic terms, I shall retain the same symbol.

In order to justify the procedure to be adopted, it is necessary to show that, if $f(t)$ be a function of t, then the rate of increase of its average value estimated over a period T, of which the beginning is variable, is equal to the average rate of its increase estimated over the same period. Now the average value of $f(t)$ estimated over the period T, beginning at the time t is $\frac{1}{\mathrm{T}}\int_t^{t+\mathrm{T}} f(t)dt$, and therefore the rate of the increase of the average value is $\frac{d}{dt}\frac{1}{\mathrm{T}}\int_t^{t+\mathrm{T}} f(t)dt$, which is equal to $\frac{1}{\mathrm{T}}\int_t^{t+\mathrm{T}} f'(t)dt$; and this last expression is the average rate of increase of $f(t)$ estimated over the same period. This therefore proves the proposition in question.

Now suppose we have $\frac{dN}{dt}=-\mathrm{M}+$ periodic terms, where M varies very slowly; then $-\mathrm{M}$ is the average value of the rate of increase of N estimated over a period which is the least common multiple of the periods of the several periodic terms. Hence by the above proposition $-\mathrm{M}$ is also the rate of increase of the average value of N estimated over the like period.

Similarly if $\frac{d\xi}{dt}=\mathrm{X}+$ periodic terms, X is the rate of increase of the average value of ξ estimated over a period, which will be the same as in the former case.

But the average value of N is the proposed new dependent variable, and the average value of ξ the new independent variable. Hence, from the present point of view, $\frac{dN}{d\xi}=-\frac{\mathrm{M}}{\mathrm{X}}$. This argument is, however, only strictly applicable, supposing there are not periodic terms in $\frac{dN}{dt}$ or $\frac{d\xi}{dt}$ of incommensurable periods, and supposing the periodic terms are rigorously circular functions, so that their amplitudes and frequencies are not functions of the time.

It is obvious, however, that if the incommensurable terms do not represent long inequalities, and if M and X vary slowly, then the theorem remains very nearly true. With respect to the variability of amplitude and frequency, it is only necessary to postulate that the so-called periodic terms are so nearly true circular functions that the integrals of them over any moderate multiple of their period is sensibly zero, to apply the argument.

Suppose, for example, $\psi(t)\cos(vt+\chi(t))$ were one of the periodic terms, then we have only to suppose that $\psi(t)$ and $\chi(t)$ vary so slowly that they remain sensibly constant during a period $\frac{2\pi}{v}$ or any moderately small multiple of it, in order to be safe in assuming $\int_0^{\frac{2\pi}{v}}\psi(t)\cos(vt+\chi(t))dt$ as sensibly zero. Now in all the inequalities in N and ξ

it is a question of days or weeks, whilst in the variations of the amplitudes and frequencies of the inequalities it is a question of millions of years. Hence the above method is safely applicable here.

It is worthy of remark that it has been nowhere assumed that the amplitudes of the periodic inequalities are small compared with the non-periodic parts of the expression.

A precisely similar argument will be applicable to every case where occasion will arise to change the independent variable. The change will accordingly be carried out without further comment, it being always understood that both dependent and independent variable are the average values of the quantities for which their symbols would in general stand.*

Then dividing (69) by (70) we have

$$-\frac{dN}{\mu d\xi}=1+\left(\frac{\tau_{\prime}}{\tau_0}\right)^2\xi^{12}+\frac{\mathrm{B}}{\mathrm{A}}\left(\frac{\tau_{\prime}}{\tau_0}\right)\xi^6 \quad . \quad . \quad . \quad . \quad . \quad . \quad . \quad . \quad . \quad (71)$$

Now $\frac{\mathrm{B}}{\mathrm{A}}=\frac{\mathrm{Z}}{\mathrm{W}\frac{\sin 4\epsilon}{\sin 2\epsilon'}+\mathrm{X}}=\sin^2 i\frac{\sin 2\epsilon'}{\sin 4\epsilon}$ approximately. This approximation will be sufficiently accurate, because the last term is small and is diminishing. For the same reason, only a small error will be incurred by treating it as constant, provided the integration be not carried over too large a field—a condition satisfied by the proposed "periods of integration." Attribute then to i, ϵ, ϵ' average values, and put

$$\beta=\tfrac{1}{13}\left(\frac{\tau}{\tau_0}\right)^2 \quad \gamma=\tfrac{1}{7}\frac{\tau_{\prime}}{\tau_0}\sin^2 i\frac{\sin 2\epsilon'}{\sin 4\epsilon} \quad . \quad . \quad . \quad . \quad . \quad . \quad . \quad . \quad . \quad (72)$$

and integrate. Then we have

$$N=1+\mu\{(1-\xi)+\beta(1-\xi^{13})+\gamma(1-\xi^7)\} \quad . \quad . \quad . \quad . \quad . \quad . \quad . \quad (73)$$

This is the approximate form of the equation of conservation of moment of momentum, and it is very nearly accurate, provided ξ does not vary too widely.

By putting $\beta=0$, $\gamma=0$, we see that the equation is independent of the obliquity, if there be only two bodies, the earth and moon, provided we neglect the fourth power of the sine of the obliquity.

The equation of reaction (68) may be written

$$\frac{dt}{d\xi}=\mu\div\frac{u^2}{\xi^{12}}(\mathrm{W}\sin 4\epsilon+\mathrm{X}\sin 2\epsilon') \quad . \quad . \quad . \quad . \quad . \quad . \quad . \quad . \quad . \quad (74)$$

* In order to feel complete confidence in my view, I placed the question before Mr. E. J. ROUTH, and with great kindness he sent me some remarks on the subject, in which he confirmed the correctness of my procedure, although he arrived at the conclusion from rather a different point of view.

Also, multiplying the equation of obliquity (66) by $\frac{dt}{d\xi}$, we have

$$\frac{\log_{10}e}{\sin i\cos i\,(1-\frac{3}{4}\sin^2 i)}\frac{di}{d\xi}=\frac{1}{N}\frac{dt}{d\xi}\left[\left(\frac{u^2}{\xi^{12}}+u_{,}^2\right)(\text{P}\sin 4\epsilon+\text{Q}\sin 2\epsilon')\right.$$
$$\left.-\frac{uu_{,}}{\xi^6}(\text{U}\sin 4\epsilon+\text{V}\sin 2\epsilon')-\frac{u^2}{\xi^{12}}\text{R}\sin 4\epsilon''\right]$$

Now by far the most important term in $\frac{d\xi}{dt}$ is that in which W occurs, and therefore $\frac{1}{2\text{W}}\frac{d\xi}{dt}$ only depends on the obliquity in its smaller term. Then, since $2\text{W}=\cos^2 i$, therefore

$$\frac{dt}{d\xi}=\frac{1}{\cos^2 i}\left(2\text{W}\frac{dt}{d\xi}\right)$$

Also

$$\frac{\cos^2 i}{\sin i\cos i\,(1-\frac{3}{4}\sin^2 i)}di=d\,.\log_e\frac{\sin i}{\sqrt{1-\frac{3}{4}\sin^2 i}}$$
$$=d\,.\log_e\tan i\,(1-\tfrac{1}{8}\sin^2 i)$$

when the fourth power of $\sin i$ is neglected.

Hence the equation may be written

$$\frac{d}{d\xi}\log_{10}\tan i(1-\tfrac{1}{8}\sin^2 i)=\frac{1}{N}\left(2\text{W}\frac{dt}{d\xi}\right)\left[\left(\frac{u^2}{\xi^{12}}+u_{,}^2\right)(\text{P}\sin 4\epsilon+\text{Q}\sin 2\epsilon')\right.$$
$$\left.-\frac{u^2}{\xi^{12}}\text{R}\sin 4\epsilon''-\frac{uu_{,}}{\xi^6}(\text{U}\sin 4\epsilon+\text{V}\sin 2\epsilon')\right]\quad\ldots\ldots\quad(75)$$

Now the term in P (which is a constant) is by far the most important of those within brackets [ ] on the right-hand side, and $2\text{W}\frac{dt}{d\xi}$ has been shown only to involve i in its smaller term. Hence the whole of the right-hand side only involves the obliquity to a subordinate degree, and, in as far as it does so, an average value may be assigned to i without producing much error.

In the equation of tidal reaction (68) or (74) also, I attribute to i in W and X an average value, and treat them as constants. As the accumulation of the error of time from period to period is unimportant, this method of approximation will give quite good enough results.

We are now in a position to track the changes in the obliquity, the day, and the month, and to find the time occupied by the changes by the method of quadratures.

First estimate an average value of i and compute Q, R . . . Z, β, γ. Take seven values of ξ, viz.: 1, ·98, ·96 . . . ·88, and calculate seven corresponding values of N; then calculate seven corresponding values of $\sin 4\epsilon$, $\sin 2\epsilon'$, $\sin 4\epsilon''$. Substitute these values in $\frac{d\xi}{dt}$, and reciprocate so as to get seven equidistant values of $\frac{dt}{d\xi}$.

Combine these seven values by WEDDLE's rule, viz. :

$$\int_0^{6h} u_x dx = \tfrac{3}{10}h[u_0+u_2+u_3+u_4+u_6+5(u_1+u_3+u_5)]$$

and so find the time corresponding to $\xi = \cdot 88$. It must be noted that the time is negative because $d\xi$ is negative.

In the course of the work the values of $\frac{dt}{d\xi}$ corresponding to $\xi=1$, $\cdot 96$, $\cdot 92$, $\cdot 88$ have been obtained. Multiply them by 2W; these values, together with the four values of $\sin 4\epsilon$, $\sin 2\epsilon'$, $\sin 4\epsilon''$ and the four of N, enable us to compute four of $\frac{d}{d\xi}\log_{10}\tan i(1-\frac{1}{8}\sin^2 i)$, as given in (75).

Combine these four values by the rule

$$\int_0^{3h} u_x dx = \frac{3h}{8}[u_0+u_3+3(u_1+u_2)]$$

and we get

$$\log_{10}\frac{\tan i(1-\frac{1}{8}\sin^2 i)}{\tan i_0(1-\frac{1}{8}\sin^2 i_0)}$$

from which the value of i corresponding to $\xi = \cdot 88$ may easily be found. It is here useless to calculate more than four values, because the function to be integrated does not vary rapidly.

We have now got final values of i, N, t corresponding to $\xi = \cdot 88$.

Since the earth is supposed to be viscous throughout the changes, therefore its figure must always be one of equilibrium, and its ellipticity of figure $e = N^2 e_0$.

Also since $\xi = \left(\frac{\Omega_0}{\Omega}\right)^{\frac{1}{3}} = \sqrt{\frac{c}{c_0}}$, where c is the moon's distance from the earth, therefore $\frac{c}{a} = \xi^2\left(\frac{c_0}{a}\right)$, which gives the moon's distance in earth's mean radii.

The fifth and sixth column of Table IV. were calculated from these formulas.

The seventh column of Table IV. shows the distribution of moment of momentum in the system; it gives μ the ratio of the moment of momentum of the moon's and earth's motion round their common centre of inertia to that of the earth's rotation round its axis, at the beginning of each period of integration.

Table I. shows the values of ϵ, ϵ', ϵ'' the angles of lagging of the semi-diurnal, diurnal, and fortnightly tides at the beginning of each period.

Tables II. and III. show the relative importance of the contributions of each term to the values of $\frac{d\xi}{dt}$ and $\frac{d}{d\xi}\log_{10}\tan i(1-\frac{1}{8}\sin^2 i)$ at the beginning of each period.

The several *lines* of the Tables II. and III. are not comparable with one another, because they are referred to different initial values of Ω and n in each line.

I will now give some details of the numerical results of each integration. The computation as originally carried out* was based on a method slightly different from that above explained, but I was able to adapt the old computation to the above method by the omission of certain terms and the application of certain correcting factors. For this reason the results in the first three tables are only given in round numbers. In the fourth table the length of day is given to the nearest five minutes, and the obliquity to the nearest five minutes of arc.

The integration begins when the length of the sidereal day is 23 hrs. 56 min., the moon's sidereal period 27·3217 m. s. days, the obliquity of the ecliptic 23° 28′, and the time zero.

First period.—Integration from $\xi=1$ to ·88; seven equidistant values computed for finding the time, and four for the obliquity.

For the obliquity the integration was not carried out exactly as above explained, in as far as that $\frac{d}{d\xi}\log_{10}\tan i$ was found instead of $\frac{d}{d\xi}\log_{10}\tan i(1-\frac{1}{8}\sin^2 i)$, but the difference in method is very unimportant. The result marked* in Table III. is $\frac{d}{d\xi}\log_{10}\tan i$.

The estimated average value of i was 22° 15′.

The final result is

$$N=1{\cdot}550,\ i=20°\ 42',\ -t=46{,}301{,}000$$

Second period.—Integration from $\xi=1$ to ·76; seven values computed for the time, and four for the obliquity.

The estimated average for i was 19°.

The final result

$$N=1{\cdot}559,\ i=17°\ 21',\ -t=10{,}275{,}000$$

Third period.—Integration from $\xi=1$ to ·76; four values computed.

The estimated average for i was 16° 30′.

The final result

$$N=1{\cdot}267,\ i=15°\ 30',\ -t=326{,}000$$

Fourth period.—Integration from $\xi=1$ to ·76; four values computed.

The estimated average for i was 15°. The small terms in β and γ were omitted in the equation of conservation of moment of momentum. All the solar and combined terms, except that in V in the equation of obliquity, were omitted.

The final result

$$N=1{\cdot}160,\ i=14°\ 25',\ -t=10{,}300$$

* I have to thank Mr. E. M. LANGLEY, of Trinity College, for carrying out the laborious computations. The work was checked throughout by myself.

TABLE I.—Showing the lagging of the several tides at the beginning of each period.

	Semi-diurnal (ϵ).	Diurnal (ϵ').	Fortnightly (ϵ'').
I.	$17\frac{1}{2}°$	$19\frac{1}{2}°$	0° 44′
II.	$23\frac{1}{2}°$	$28\frac{1}{2}°$	1° 5′
III.	$29\frac{1}{2}°$	40°	2° 27′
IV.	$32\frac{1}{2}°$	$46\frac{1}{2}°$	5° 30′

TABLE II.—Showing the contribution of the several tidal effects to tidal reaction $\left(i.e., \text{ to } \frac{d\xi}{dt}\right)$ at the beginning of each period. The numbers to be divided by 10^{10}.

	Semi-diurnal.	Diurnal.
I.	12·	1·2
II.	69·	6·3
III.	2200·	200·
IV.	70000·	6100·

TABLE III.—Showing the contributions of the several tidal effects to the change of obliquity (*i.e.*, to $\frac{d}{d\xi}\log_{10}\tan i(1-\frac{1}{8}\sin^2 i)$) at the beginning of each period.

	Lunar semi-diurnal.	Lunar diurnal.	Solar semi-diurnal.	Solar diurnal.	Combined semi-diurnal.	Combined diurnal.	Fortnightly.	$\frac{d}{d\xi}\log\tan i\,(1-\frac{1}{8}\sin^2 i)$.
*I.	·82	·13	·18	·03	−·06	−·48	−·006	·60*
II.	·44	·06	·02	..	−·01	−·16	−·003	·36
III.	·22	·03	..	..	..	−·02	−·003	·23
IV.	·13	·02	..	..	..	..	−·004	·14

TABLE IV.—Showing the physical meaning of the results of the integration.

	Time ($-t$).	Sidereal day in m. s. hours.	Moon's sidereal period in m. s. days.	Obliquity of ecliptic (i).	Reciprocal of ellipticity of figure.	Moon's distance in earth's mean radii.	Ratio of m. of m. of orbital motion to m. of m. of earth's rotation.	Heat generated (see Section 16).
	Years.	h. m.	d.					Degrees Fahr.
Initial state.	0	23 56	27·32	23° 28′	232	60·4	4·01	0°
I.	46,300,000	15 30	18·62	20° 40′	96	46·8	2·28	225°
II.	56,600,000	9 55	8·17	17° 20′	40	27·0	1·11	760°
III.	56,800,000	7 50	3·59	15° 30′*	25	15·6	·67	1300°
IV.	56,810,000	6 45	1·58	14° 25′*	18	9·0	·44	1760°

The whole of these results are based on the supposition that the plane of the lunar orbit will remain very nearly coincident with the ecliptic throughout these changes. I now (July, 1879), however, see reason to believe that the secular changes in the plane of the lunar orbit will have an important influence on the obliquity of the ecliptic. Up to the end of the second period the change of obliquity as given in Table IV. will be approximately correct, but I find that during the third and fourth periods of integration there will be a phase of considerable nutation. The results in the column of obliquity marked (*) have not, therefore, very much value as far as regards the explanation of the obliquity of the ecliptic; they are, however, retained as being instructive from a dynamical point of view.

§ 16. *The loss of energy of the system.*

It is obvious that as there is tidal friction the moon-earth system must be losing energy, and I shall now examine how much of this lost energy turns into heat in the interior of the earth. The expressions potential and kinetic energy will be abbreviated by writing them *p.e.* and *k.e.*

The *k.e.* of the earth's rotation is $\frac{1}{5}Ma^2n^2$.

The *k.e.* of the earth's and moon's orbital motion round their common centre of inertia is

$$\tfrac{1}{2}M\left(\frac{m\mathrm{r}}{m+M}\right)^2\Omega^2+\tfrac{1}{2}m\left(\frac{M\mathrm{r}}{m+M}\right)^2\Omega^2=\tfrac{1}{2}M\mathrm{r}^2\frac{\Omega^2}{1+\nu}.$$

But since the moon's orbit is circular $\Omega^2\mathrm{r}=g\left(\frac{a}{\mathrm{r}}\right)^2\frac{1+\nu}{\nu}$, so that $\frac{\Omega^2\mathrm{r}^2}{1+\nu}=\frac{ga^2}{\nu\mathrm{r}}$. Hence the whole *k.e.* of the moon-earth system is

$$Ma^2\left(\tfrac{1}{5}n^2+\tfrac{1}{2}\frac{g}{\nu}\frac{1}{\mathrm{r}}\right)$$

The $p.e.$ of the system is

$$-\frac{Mm}{\mathrm{r}}=-\frac{M}{\nu}\frac{ga^2}{\mathrm{r}}$$

Therefore the whole energy E of the system is

$$Ma^2\left\{\tfrac{1}{5}n^2-\frac{1}{2\nu}\frac{g}{\mathrm{r}}\right\}$$

and in gravitation units

$$E=Ma\left\{\tfrac{1}{5}\frac{n^2a}{g}-\frac{1}{2\nu}\frac{a}{\mathrm{r}}\right\}$$

Now since the earth is supposed to be plastic throughout all these changes, therefore its ellipticity of figure

$$e=\tfrac{5}{4}\frac{n^2a}{g}$$

and

$$E=Ma\left\{\tfrac{4}{25}e-\frac{1}{2\nu}\frac{a}{\mathrm{r}}\right\}$$

If e, $e+\Delta e$ and r, r$+\Delta$r be the ellipticity of figure, and the moon's distance at two epochs, if J be JOULE's equivalent, and σ the specific heat of the matter constituting the earth; then the loss of energy of the system between these two epochs is sufficient to heat unit mass of the matter constituting the earth

$$-\frac{Ma}{\mathrm{J}\sigma}\left\{\tfrac{4}{25}\Delta e-\frac{1}{2\nu}\Delta\frac{a}{\mathrm{r}}\right\} \text{ degrees,}$$

and is therefore enough to heat the whole mass of the earth

$$-\frac{a}{\mathrm{J}\sigma}\left\{\tfrac{4}{25}\Delta e-\frac{1}{2\nu}\Delta\frac{a}{\mathrm{r}}\right\} \text{ degrees.}$$

It must be observed that in this formula the whole loss of $k.e.$ of the earth's rotation, due both to solar and lunar tidal friction, is included, whilst only the gain of the moon's $p.e.$ is included, and the effect of the solar tidal reaction in giving the earth greater potential energy relatively to the sun is neglected.

In the fifth and sixth columns of Table IV. of the last section the ellipticity of figure and the moon's distance in earth's radii are given; and these numbers were used in calculating the eighth column of the same table.

I used British units, so that 772 foot-pounds being required to heat 1 lb. of water 1° Fahr., J=772; the specific heat of the earth was taken as $\frac{1}{6}$th, which is about that of iron, many of the other metals having a still smaller specific heat; the earth's radius was taken, as before, equal to 20·9 million feet. Then the last column states that energy enough has been turned into heat in the interior of the earth to warm its whole mass so many degrees Fahrenheit within the times given in the first column of the same table.

The consideration of the distribution of the generation of heat and the distortion of the interior of the earth must be postponed to a future occasion.

In the succeeding paper I have considered the bearing of these results on the secular cooling of the earth, and in a subsequent paper ('Proceedings of the Royal Society,' No. 197, June 19, 1879, p. 168) the general problem of tidal friction is considered by the aid of the theory of energy.

§ 17. *Integration in the case of small variable viscosity.**

In the solution of the problem which has just been given, where the viscosity is constant, the obliquity of the ecliptic does not diminish as fast as it might do as we look backwards. The reason of this is that the ratio of the negative terms to the positive ones in the equation of obliquity is not as small as it might be; that ratio principally depends on the fraction $\frac{\sin 2\epsilon'}{\sin 4\epsilon}$, which has its smallest value when ϵ is very small.

I shall now, therefore, consider the case where the viscosity is small, and where it so varies that ϵ always remains small.

This kind of change of viscosity is in general accordance with what one may suppose to have been the case, if the earth was a cooling body, gradually freezing as it cooled.

The preceding solution is moreover somewhat unsatisfactory, inasmuch as the three semi-diurnal tides are throughout supposed to suffer the same retardation, as also are the three diurnal tides; and this approximation ceases to be sufficiently accurate towards the end of the integration.

In the present solution the retardations of all the lunar tides will be kept distinct.

By (40) and (40′), Section 11,

$$\tan 2\epsilon_1=\frac{2(n-\Omega)}{\rho},\ \tan 2\epsilon=\frac{2n}{\rho},\ \tan 2\epsilon_2=\frac{2(n+\Omega)}{\rho},\ \tan 2\epsilon''=\frac{2\Omega}{\rho},$$

$$\tan \epsilon'_1=\frac{n-2\Omega}{\rho},\quad \tan \epsilon'=\frac{n}{\rho},\quad \tan \epsilon'_2=\frac{n+2\Omega}{\rho}$$

for the lunar tides.

For the solar tides we may safely neglect Ω, compared with n, and we have

* This section has been partly rewritten and rearranged, and wholly recomputed since the paper was presented. The alterations are in the main dated December 19, 1878.

$\tan 2\epsilon = \frac{2n}{\rho}$, $\tan \epsilon' = \frac{n}{\rho}$ for the semi-diurnal and diurnal tides respectively. The semi-annual tide will be neglected.

Then if the viscosity so varies that all the ϵ's are always small, and if we put $\frac{\Omega}{n} = \lambda$, we have

$$\left.\begin{array}{lll} \frac{\sin 4\epsilon_1}{\sin 4\epsilon} = 1-\lambda, & \frac{\sin 4\epsilon_2}{\sin 4\epsilon} = 1+\lambda, & \frac{\sin 4\epsilon''}{\sin 4\epsilon} = \lambda \\ \frac{\sin 2\epsilon'_1}{\sin 4\epsilon} = \frac{1}{2}-\lambda, & \frac{\sin 2\epsilon'}{\sin 4\epsilon} = \frac{1}{2}, & \frac{\sin 2\epsilon'_2}{\sin 4\epsilon} = \frac{1}{2}+\lambda \end{array}\right\} \quad . \quad . \quad . \quad . \quad . \quad . \quad (76)$$

By means of these equations we may expresss all the sines of the ϵ's in terms of $\sin 4\epsilon$.

Then, remembering that the spheroid is viscous, and that therefore $E_1 = \cos 2\epsilon_1$, $E'_1 = \cos \epsilon'_1$, &c., we have by Sections 4 and 7, equations (16) and (29),

$$\frac{di_{m^2}}{dt} = \frac{1}{N}\frac{\tau^2}{\mathfrak{g}n_0}\Big[\tfrac{1}{2}p^7q \sin 4\epsilon_1 - p^3q^3(p^2-q^2)\sin 4\epsilon - \tfrac{1}{2}pq^7 \sin 4\epsilon_2 - \tfrac{3}{2}p^3q^3 \sin 4\epsilon''$$
$$+\tfrac{1}{2}p^5q(p^2+3q^2)\sin 2\epsilon'_1 - \tfrac{1}{2}pq(p^2-q^2)^3 \sin 2\epsilon' - \tfrac{1}{2}pq^5(3p^2+q^2)\sin 2\epsilon'_2\Big] \quad . \quad (77)$$

$$-\frac{dN_{m^2}}{dt} = \frac{\tau^2}{\mathfrak{g}n_0}\Big[\tfrac{1}{2}p^8 \sin 4\epsilon_1 + 2p^4q^4 \sin 4\epsilon + \tfrac{1}{2}q^8 \sin 4\epsilon_2$$
$$+p^6q^2 \sin 2\epsilon'_1 + p^2q^2(p^2-q^2)^2 \sin 2\epsilon' + p^2q^6 \sin 2\epsilon'_2\Big] \quad . \quad (78)$$

And by (57), Section 14,

$$\mu\frac{d\xi}{dt} = \frac{\tau^2}{\mathfrak{g}n_0}\Big[\tfrac{1}{2}p^8 \sin 4\epsilon_1 - \tfrac{1}{2}q^8 \sin 4\epsilon_2 - 3p^4q^4 \sin 4\epsilon'' + 2p^6q^2 \sin 2\epsilon'_1 - 2p^2q^6 \sin 2\epsilon'_2\Big] \quad . \quad (79)$$

The first two of these equations only refer to the action of the moon on the lunar tides; but the last is the same whether there be solar tides or not.

Then if we substitute from (76) for all the ϵ's in terms of $\sin 4\epsilon$, and introduce $\cos i = P = p^2 - q^2$, $\sin i = Q = 2pq$, we find on reduction

$$\left.\begin{array}{l} \frac{di_{m^2}}{dt} = \frac{1}{N}\frac{\tau^2}{\mathfrak{g}n_0}\sin 4\epsilon\big[\tfrac{1}{4}PQ - \tfrac{1}{2}\lambda Q\big] \\ -\frac{dN_{m^2}}{dt} = \frac{1}{2}\frac{\tau^2}{\mathfrak{g}n_0}\sin 4\epsilon\big[1 - \tfrac{1}{2}Q^2 - \lambda P\big] \\ \mu\frac{d\xi}{dt} = \frac{1}{2}\frac{\tau^2}{\mathfrak{g}n_0}\sin 4\epsilon\big[P - \lambda\big] \end{array}\right\} \quad . \quad . \quad . \quad . \quad . \quad . \quad . \quad (80)$$

The parts of $\frac{di}{dt}$ and $\frac{dN}{dt}$ which arise from the attraction of the sun on the solar tides may be at once written down by symmetry, and $\lambda_{,}=\frac{\Omega_{,}}{n}$ may be considered as a small fraction to be neglected compared with unity. Thus we have

$$\left.\begin{aligned}\frac{di_{m_{,}^{2}}}{dt}&=\frac{1}{N}\frac{\tau_{,}^{2}}{\mathfrak{g}n_0}\sin 4\epsilon.\tfrac{1}{4}PQ\\ -\frac{dN_{m_{,}^{2}}}{dt}&=\tfrac{1}{2}\frac{\tau_{,}^{2}}{\mathfrak{g}n_0}\sin 4\epsilon(1-\tfrac{1}{2}Q^2)\end{aligned}\right\} \quad . \quad (81)$$

Lastly as to the terms due to the combined action of the two disturbing bodies, it was remarked that they only involved ϵ and ϵ', which are independent of the orbital motions.

Thus by (33) we have

$$\left.\begin{aligned}\frac{di_{mm_{,}}}{dt}&=-\frac{1}{N}\frac{\tau\tau_{,}}{\mathfrak{g}n_0}\sin 4\epsilon.\tfrac{1}{4}PQ\\ -\frac{dN_{mm_{,}}}{dt}&=\frac{\tau\tau_{,}}{\mathfrak{g}n_0}\sin 4\epsilon.\tfrac{1}{4}Q^2\end{aligned}\right\} \quad . \quad . \quad (82)$$

Then collecting results from the last three sets of equations and substituting $\cos i$ and $\sin i$ for P and Q, and $\frac{\Omega}{n}$ for λ, we have

$$\left.\begin{aligned}\frac{di}{dt}&=\frac{1}{N}\frac{\sin 4\epsilon}{\mathfrak{g}n_0}\tfrac{1}{4}\sin i\cos i\left[\tau^2+\tau_{,}^{2}-\tau\tau_{,}-\frac{2\Omega}{n}\tau^2\sec i\right]\\ -\frac{dN}{dt}&=\tfrac{1}{2}\frac{\sin 4\epsilon}{\mathfrak{g}n_0}\left[(1-\tfrac{1}{2}\sin^2 i)(\tau^2+\tau_{,}^{2})+\tfrac{1}{2}\tau\tau_{,}\sin^2 i-\frac{\Omega}{n}\tau^2\cos i\right]\\ \mu\frac{d\xi}{dt}&=\tfrac{1}{2}\frac{\sin 4\epsilon}{\mathfrak{g}n_0}\cos i\,\tau^2\left(1-\frac{\Omega}{n}\sec i\right)\end{aligned}\right\} \quad . \quad . \quad (83)$$

These are the simultaneous equations which are to be solved.

Subject to the special hypothesis regarding the relationship between the retardations of the several tides, and except for the neglect of a term $-\frac{2\Omega_{,}}{n}\tau_{,}^{2}\sec i$ in the first of them, and of $-\frac{\Omega_{,}}{n}\tau_{,}^{2}\cos i$ in the second, they are rigorously true.

We will first change the independent variable in the first two equations from t to ξ.

Dividing the first and second equations by the third, and observing that

$$\frac{2di}{\sin i}=d\log\tan^2\frac{i}{2}$$

we have

$$\left.\begin{aligned}\frac{d}{\mu d\xi}\log\tan^2\frac{i}{2}&=\frac{1+\left(\frac{\tau_{\prime}}{\tau}\right)^2-\left(\frac{\tau_{\prime}}{\tau}\right)-\frac{2\Omega}{n}\sec i}{N\left(1-\frac{\Omega}{n}\sec i\right)}\\ -\frac{dN}{\mu d\xi}&=\frac{\frac{1-\frac{1}{2}\sin^2 i}{\cos i}\left[1+\left(\frac{\tau_{\prime}}{\tau}\right)^2\right]+\frac{1}{2}\left(\frac{\tau_{\prime}}{\tau}\right)\sin i\tan i-\frac{\Omega}{n}}{1-\frac{\Omega}{n}\sec i}\end{aligned}\right\}\quad . \; . \; . \; . \quad (84)$$

If there be only one disturbing body, which is an interesting case from a theoretical point of view, the equations may be found by putting $\tau_{\prime}=0$, and may then be written

$$\left.\begin{aligned}\frac{d}{\mu d\xi}\log\tan^2\frac{i}{2}&=\frac{1}{N}\frac{\cos i-\frac{2\Omega}{n}}{\cos i-\frac{\Omega}{n}}\\ -\frac{dN}{\mu d\xi}&=\frac{1-\frac{1}{2}\sin^2 i-\frac{\Omega}{n}\cos i}{\cos i-\frac{\Omega}{n}}\\ \mu\frac{d\xi}{dt}&=\tfrac{1}{2}\sin 4\epsilon.\frac{\tau^2}{gn_0}\left(\cos i-\frac{\Omega}{n}\right)\end{aligned}\right\}\quad . \; . \; . \; . \; . \; . \; . \; . \; . \; . \quad (85)$$

From these equations we see that so long as Ω is less than $n\cos i$, the satellite recedes from the planet as the time increases, and the planet's rotation diminishes, because the numerator of the second equation may be written $\cos i\left(\cos i-\frac{\Omega}{n}\right)+\frac{1}{2}\sin^2 i$, which is essentially positive so long as Ω is less than $n\cos i$. But the tidal friction vanishes whenever $\Omega=n\frac{1+\cos^2 i}{2\cos i}$. The fraction $\frac{1+\cos^2 i}{2\cos i}$ is however necessarily greater than unity, and therefore the tidal friction cannot vanish, unless the month be as short or shorter than the day. The obliquity increases if Ω be less than $\frac{1}{2}n\cos i$, but diminishes if it be greater than $\frac{1}{2}n\cos i$. Hence the equation $\Omega=\frac{1}{2}n\cos i$ gives the relationship which determines the position and configuration of the system for instantaneous dynamical stability with regard to the obliquity (compare the figures 2, 3, 4, Plate 36). From this it follows that the position of zero obliquity is one of dynamical stability for all values of n between Ω and 2Ω, but if n be greater than 2Ω, this position is unstable.*

We will now return to the problem regarding the earth. We may here regard $\frac{\Omega}{n}$ as a small fraction, and i as sufficiently small to permit us to neglect $\frac{1}{8}\sin^4 i$; also $\left(\frac{\Omega}{n}\sec i\right)^2$, $\frac{\tau_{\prime}}{\tau}\frac{\Omega}{n}\sec i$, $\left(\frac{\tau_{\prime}}{\tau}\right)^2\frac{\Omega}{n}\sec i$ will be neglected.

* Added on September 25, 1879.—The result in the text applies to the case of evanescent viscosity. If the viscosity be infinitely large the sines of twice the angles of lagging will be inversely instead of directly proportional to the speeds of the corresponding tides (compare p. 482). Thus we must here invert the right-hand sides of the six equations (76). If the obliquity be very small (77), (78), (79) become

$$\left.\begin{aligned}\frac{di}{dt}&=\frac{1}{N}\frac{\tau^2}{\mathfrak{g}n_0}\tfrac{1}{4}\sin i\sin 4\epsilon_1\left[1+\frac{2(1-\lambda)}{1-2\lambda}-2(1-\lambda)\right]\\&=\frac{1}{N}\frac{\tau^2}{\mathfrak{g}n_0}\tfrac{1}{4}\sin i\sin 4\epsilon_1\left(\frac{1+2\lambda-4\lambda^2}{1-2\lambda}\right)\\-\frac{dN}{dt}&=\mu\frac{d\xi}{dt}=\frac{\tau^2}{\mathfrak{g}n_0}\tfrac{1}{2}\sin 4\epsilon_1\end{aligned}\right\} \quad . \quad . \quad . \quad . \quad . \quad . \quad . \quad (85')$$

When $2\lambda=1$, $\frac{di}{dt}$ apparently becomes infinite; but in this case the viscosity must be infinitely large in order to make the tide of speed $n-2\Omega$ lag at all, and if it be infinitely large $\sin 4\epsilon_1$ is infinitely small. If the viscosity be large but finite, then when $2\lambda=1$, the slow diurnal tide of speed $n-2\Omega$ is no longer a true tide, but is a permanent alteration of figure of the spheroid. Thus $\epsilon'_1=0$ and $\frac{di}{dt}$ depends on $[\sin 4\epsilon_1-\sin 2\epsilon']$ which is equal to $\sin 4\epsilon_1[1-2(1-\lambda)]$ when the viscosity is large, and vanishes when $2\lambda=1$. Thus when the viscosity is very large (not infinite) $\frac{di}{dt}$ vanishes when $2\Omega\div n=1$, as it does when the viscosity is very small.

When $1+2\lambda-4\lambda^2=0$, that is, when $\lambda=\frac{\sqrt{5}+1}{4}=1\div 1{\cdot}236$, $\frac{di}{dt}$ vanishes; and it is negative if λ be a little greater, and positive if a little less than $1\div 1{\cdot}236$. And $1-2\lambda$ is negative if λ be greater than $\frac{1}{2}$.

Hence it follows that *for large viscosity of the planet, zero obliquity is dynamically unstable, if the satellite's period be less than* 1·236 *of the planet's period of rotation; is stable if the satellite's period be between* 1·236 *and* 2 *of the planet's period; and is unstable for longer periods of the satellite.*

If the viscosity be very large $\frac{N}{\mu}\frac{d}{d\xi}\log\tan^2\frac{i}{2}=\frac{1+2\lambda-4\lambda^2}{1-2\lambda}$, but if the viscosity be very small the same expression $=\frac{1-2\lambda}{1-\lambda}$. For positive values of λ, less than 1 and greater than ·6910 or $1\div 1{\cdot}447$, the former is less than the latter, and if λ be less than $1\div 1{\cdot}447$ and greater than 0 the former is greater than the latter.

Hence if there be only a single satellite, as soon as the month is longer than two days, the obliquity of the planet's axis to the plane of the satellite's orbit will increase more, in the course of evolution, for large than for small viscosities. This result is reversed if there be two satellites, as we see by comparing figs. 2 and 4, Plate 36.

Then our equations are

$$\left.\begin{aligned}\frac{d}{\mu d\xi}\log_e \tan^2\frac{i}{2}&=\frac{1+\left(\frac{\tau_{\prime}}{\tau}\right)^2-\left(\frac{\tau_{\prime}}{\tau}\right)-\frac{2\Omega}{n}\sec i}{N\left(1-\frac{\Omega}{n}\sec i\right)}\\ -\frac{dN}{\mu d\xi}&=1+\left(\frac{\tau_{\prime}}{\tau}\right)^2+\frac{1}{2}\frac{\tau_{\prime}}{\tau}\sin i\tan i+\frac{\Omega}{n}(\sec i-1)\end{aligned}\right\}\quad . \quad . \quad . \quad . \quad . \quad . \quad (86)$$

The experience of the preceding integration shows that i varies very slowly compared with the other variables N and ξ; hence in integrating these equations an average value will be attributed to i, as it occurs in small terms on the right-hand sides of these equations.

The second equation will be considered first.

We have $\tau=\frac{\tau_0}{\xi^6}$, so that if we put $\beta=\frac{1}{13}\left(\frac{\tau_{\prime}}{\tau_0}\right)^2$, $\gamma=\frac{1}{14}\frac{\tau_{\prime}}{\tau_0}\sin i\tan i$, and omit the last term, we get by integrating from 1 to N and from 1 to ξ

$$N=1+\mu\{1-\xi+\beta(1-\xi^{13})+\gamma(1-\xi^7)\}\quad . \quad . \quad . \quad . \quad . \quad . \quad . \quad (87)$$

as a first approximation. This is the form which was used in the previous solution, for, by classifying the tides in three groups as regards retardation of phase, we virtually neglected Ω compared with n.

This equation will be sufficiently accurate so long as $\frac{\Omega}{n}$ is a moderately small fraction; but we may obtain a second approximation by taking account of the last term.

Now

$$\begin{aligned}\frac{\Omega}{n}(\sec i-1)&=\tfrac{1}{2}\sin^2 i\,\frac{\Omega_0}{n_0}\cdot\frac{1}{N\xi^3}\text{ very nearly}\\ &=\tfrac{1}{2}\sin^2 i\,\frac{\Omega_0}{\mu n_0}\cdot\frac{1}{\xi^3\left[\frac{1+\mu}{\mu}-\xi\right]}\end{aligned}$$

by substituting an approximate value for N.

A more correct form for the equation of conservation of moment of momentum will be given by adding to the right-hand side of equation (87) the integral of this last expression from 1 to ξ and multiplying it by μ. And in effecting this integration i may be regarded as constant.

Let $k=\frac{1+\mu}{\mu}$. Then since

$$\frac{1}{\xi^3(k-\xi)}=\frac{1}{k\xi^3}+\frac{1}{k^2\xi^2}+\frac{1}{k^3\xi}+\frac{1}{k^3}\cdot\frac{1}{(k-\xi)}$$

Therefore

$$\int_\xi^1 \frac{d\xi}{\xi^3(k-\xi)}=\frac{1}{2k}\left(\frac{1}{\xi^2}-1\right)+\frac{1}{k^2}\left(\frac{1}{\xi}-1\right)+\frac{1}{k^3}\log\frac{k-\xi}{\xi(k-1)}$$

$$=\frac{\mu}{2(1+\mu)}\left(\frac{1}{\xi}-1\right)\left(\frac{1}{\xi}+\frac{1+3\mu}{1+\mu}\right)+\left(\frac{\mu}{\mu+1}\right)^3\log\left[\frac{1+\mu(1-\xi)}{\xi}\right]$$

Hence the second approximation is

$$N=1+\mu\{(1-\xi)+\beta(1-\xi^{13})+\gamma(1-\xi^7)\}+\tfrac{1}{4}\sin^2 i\frac{\Omega_0}{n_0}\frac{\mu}{1+\mu}\left(\frac{1}{\xi}-1\right)\left(\frac{1}{\xi}+\frac{1+3\mu}{1+\mu}\right)$$
$$+\tfrac{1}{2}\sin^2 i\frac{\Omega_0}{n_0}\left(\frac{\mu}{\mu+1}\right)^3\log\left[\frac{1+\mu(1-\xi)}{\xi}\right]. \quad . \quad (88)$$

It would no doubt be possible to substitute this approximate value of N in terms of ξ, in the equation which gives the rate of change of obliquity, and then to find an approximate analytical integral of the first equation. But the integral would be very long and complicated, and I prefer to determine the amount of change of obliquity by the method of quadratures.

In the present case it is obviously useless to try to obtain the time occupied by the changes, without making some hypothesis with regard to the law governing the variations of viscosity; and even supposing the viscosity small but constant during the integration, the time would vary inversely as the coefficient of viscosity, and would thus be arbitrary. The only thing which can be asserted is that if the viscosity be small, the changes proceed more slowly than in the case which has been already solved numerically.

To return, then, to the proposed integration by quadratures: by means of the equation (88) we may compute four values of N (corresponding, say, to $\xi=1$, ·96, ·92, ·88); and since $\tau=\frac{\tau_0}{\xi^6}$, and $\frac{\Omega}{n}=\frac{\Omega_0}{n_0}\frac{1}{N\xi^3}$, we may compute four equidistant values of all the terms on the right-hand side of the first of equations (86), except in as far as i is involved. Now i being only involved in small terms, we may take as an approximate final value of i that which is given by the solution of Section 15, and take as the four corresponding values i_0, $i_0+\frac{i-i_0}{3}$, $i_0+2\frac{(i-i_0)}{3}$, i.

Hence four equidistant values of the right-hand side may be computed, and combined by the rule $\int_0^{3h}u_x dx=\frac{3h}{8}[u_0+u_3+3(u_1+u_2)]$, which will give the integral of the right-hand side from ξ to 1; and this is equal to $\log\tan^2\frac{i}{2}-\log\tan^2\frac{i_0}{2}$.

The integration was divided into a number of periods, just as in the solution of Section 15. The following were the results :

First period. From $\xi=1$ to $\cdot 88$; $\mu=4\cdot 0074$; $i=20^\circ\ 28'$; $N=1\cdot 5478$. The term in $\frac{\Omega_0}{n_0}$ in the expression for N added $\cdot 0012$ to the value of N.

Second period. From $\xi=1$ to $\cdot 76$; $\mu=2\cdot 2784$; $i=17^\circ\ 4'$; $N=1\cdot 5590$. The term in $\frac{\Omega_0}{n_0}$ added $\cdot 0011$ to the value of N.

Third period. From $\xi=1$ to $\cdot 76$; $\mu=1\cdot 1107$; $i=15^\circ\ 22'$; $N=1\cdot 2677$. The term in $\frac{\Omega_0}{n_0}$ added $\cdot 0007$ to the value of N.

It may be observed that during the first period of integration $\frac{\Omega}{n}$ diminishes, and reaches its minimum about the end of the period. During the rest of the integration it increases. If we neglect the solar action and the obliquity, it is easy to find the minimum value of $\frac{\Omega}{n}$. For $\frac{\Omega}{n}=\frac{\Omega_0}{n_0}\frac{1}{N\xi^3}$ and reaches its minimum when $\frac{dN}{d\xi}=-\frac{3N}{\xi}$; but $\frac{dN}{d\xi}=-\mu$. Therefore $N=\frac{\xi\mu}{3}$. Now $N=1+\mu(1-\xi)$, and hence $\xi=\frac{3}{4}\frac{1+\mu}{\mu}$. If $\mu=4$, $\xi=\frac{15}{16}=\cdot 9375$. This value of ξ is passed through at near the end of the first period of integration. At this period there are $19\cdot 2$ mean solar hours in the day; $22\frac{1}{2}$ mean solar days in the sidereal month; and $28\frac{1}{7}$ rotations of the earth in the sidereal month. This result of $28\frac{1}{7}$ is, of course, only approximate, the true result being about 29.*

The physical meaning of these results is given in a table below.

At the end of the third period of integration the solar terms (those in $\frac{\tau_{\prime}}{\tau}$) have become small in all the equations, and as they are rapidly diminishing they may be safely neglected. To continue the integration from this point a slight variation of method will be convenient.

Our equations may now be written approximately

$$N=1+\mu(1-\xi)$$

$$-\frac{d}{dN}\log\tan^2\frac{i}{2}=\frac{1}{N}\frac{1-\frac{2\Omega}{n}\sec i}{1-\frac{\Omega}{n}}$$

In order to find how large a diminution of obliquity is possible if the integration be continued, we require to stop at the point where $n\cos i=2\Omega$.

Now the equation $N=1+\mu(1-\xi)$ may be written

$$\frac{n}{n_0}=1+\mu\left(1-\sqrt[3]{\frac{\Omega_0}{\Omega}}\right).$$

* The subject is referred to from a more general point of view in a paper on the "Secular Effects of Tidal Friction," see 'Proc. Roy. Soc.,' No. 197, 1879.

If therefore we put $x=\sqrt[3]{\Omega}$, we must stop the integration at the point where $n=2x^3 \sec i$, x being given by the equation

$$\frac{2x^3 \sec i}{n_0}=1+\mu\left[1-\frac{\sqrt[3]{\Omega_0}}{x}\right]$$

And if we assume $i=14°$, x is given by

$$x^4-\tfrac{1}{2}n_0 \cos 14°(1+\mu)x+\frac{1}{2s}\cos 14°=0$$

because $\mu=1\div sn_0\Omega_0^{\frac{1}{3}}$.

Now at the end of the third period of integration, which is the beginning of the new period, I found

$$\log n_0=3{\cdot}84753,\ \log \mu=9{\cdot}82338-10, \text{ and } \log s=5{\cdot}39378-10$$

The unit of time being the present tropical year.

Hence the equation is

$$x^4-5690x+19586=0$$

The required root is nearly $\sqrt[3]{5690}$, and a second approximation gives $x=\Omega^{\frac{1}{3}}=16{\cdot}703$ (16·51 would have been more accurate).

But $\Omega_0^{\frac{1}{3}}=8{\cdot}616$. Hence we desire to stop the integration when

$$\xi=\left(\frac{\Omega_0}{\Omega}\right)^{\frac{1}{3}}=\frac{8{\cdot}616}{16{\cdot}703}={\cdot}516.$$

Now $\mu={\cdot}6659$; hence when $\xi={\cdot}516$, $N=1{\cdot}322$.

In order to integrate the equation of obliquity by quadratures, I assume the four equidistant values,

$$N=1{\cdot}000, \quad 1{\cdot}107, \quad 1{\cdot}214, \quad 1{\cdot}321$$

And by means of the equation $\xi=1-\frac{N-1}{{\cdot}6659}=1-(N-1)(1{\cdot}502)$ the corresponding values of ξ are found to be

$$1{\cdot}000, \quad {\cdot}8393, \quad {\cdot}6786, \quad {\cdot}5179$$

Then by means of the formula $\frac{\Omega}{n}=\frac{\Omega_0}{n_0}\frac{1}{N\xi^3}$, the corresponding values of $\frac{\Omega}{n}$ are found to be

$${\cdot}0909, \quad {\cdot}1388, \quad {\cdot}2395, \quad {\cdot}4951$$

I assumed conjecturally four values of i lying between $i_0=15° \, 22'$ and $i=14°$, which I knew would be very nearly the final value of i; and then computed four equidistant values of $-\frac{d}{dN}\log_{10}\tan\frac{i}{2}$.

The values were

$${\cdot}19381, \quad {\cdot}16230, \quad {\cdot}11882, \quad -{\cdot}00684.$$

The fact that the last value is negative shows that the integration is carried a little beyond the point when $n \cos i=2\Omega$, but this is unimportant.

Combining these values by the rules of the calculus of finite differences, I find $i=13° 59'$.

This final value of ξ (viz.: ·5179) makes the moon's sidereal period 12 hours, and the value of N (viz.: 1·321) makes the day 5 hours 55 minutes.

These results complete the integration of the fifth period.

The physical meaning of the results for all five periods is given in the following table:—

Sidereal day in m.s. hours and minutes.	Moon's sidereal period in m.s. days.	Obliquity of ecliptic.
h. m.		
Initial 23 56	27·32 days	23° 28′
15 28	18·62 „	20° 28′
9 55	8·17 „	17° 4′
7 49	3·59 „	15° 22′ *
Final 5 55	12 hours	14° 0′ *

It is worthy of notice that at the end of the first period there were 28·9 days of that time in the then sidereal month; whilst at the end of the second period there were only 19·7. It seems then that at the present time tidal friction has, in a sense, done more than half its work, and that the number of days in the month has passed its maximum on its way towards the state of things in which the day and month are of equal length—as investigated in the following section.

In the last column of the preceding table the last two results in the column giving the obliquity of the ecliptic (which are marked with asterisks) cannot safely be accepted, because, as I have reason to believe, the simultaneous changes of inclination of the lunar orbit will, after the end of the second period of integration, have begun to influence the results perceptibly.

For this same reason the integration, which has been carried to the critical point where $n \cos i=2\Omega$, and where $\frac{di}{dt}$ changes sign, will not be pursued any further. Nevertheless we shall be able to trace the moon's periodic time, and the length of day to their initial condition. It is obvious that as long as n is greater than Ω, there will be tidal friction, and n will continue to approach Ω, whilst both increase retrospectively in magnitude.

I shall now refer to a critical phase in the relationship between n and Ω, of a totally different character from the preceding one, and which must occur at a point a little more remote in time than that at which the above integration stops.

This critical phase occurs when the free nutation of the oblate spheroid has a frequency equal to that of the forced fortnightly nutation.

In the ordinary theory of the precession and nutation of a rigid oblate spheroid, the fortnightly nutation arises out of terms in the couples acting about a pair of axes fixed in the equator, which have speeds $n-2\Omega$ and $n+2\Omega$. If C and A be the greatest and least principal moments of inertia, then on integration these terms are divided by $\frac{C-A}{A}n+n\mp2\Omega$ and give rise to terms in $\frac{di}{dt}$ and $\frac{d\psi}{dt}\sin i$ of speed 2Ω. When 2Ω is neglected compared with n, we obtain the formula, given in any work on physical astronomy, for the fortnightly nutation.

Now it is obvious that if $\frac{C-A}{A}n+n=2\Omega$, the former of these two terms becomes infinite. Since in our case the spheroid in homogeneous $\frac{C-A}{A}=e$ the ellipticity of the spheroid; and since the spheroid is viscous $e=\frac{1}{2}\frac{n^2}{\mathfrak{g}}$. Therefore the critical relationship is $\frac{1}{2}\frac{n^3}{\mathfrak{g}}+n=2\Omega$.

When this condition is satisfied the ordinary solution is nugatory, and the true solution represents a nutation the amplitude of which increases with the time.

The critical point where the above integration stops is given by $\frac{2\Omega}{n}=\cos i$, and this critical point by $\frac{2\Omega}{n}=1+\frac{1}{2}\frac{n^2}{\mathfrak{g}}$; it follows therefore that $\frac{\Omega}{n}$ is little larger in the second case than in the first. Therefore this critical point has not been already reached where the integration stops, but will occur shortly afterwards.

It is obvious that the amplitude of the nutation cannot increase for an indefinite time, because the critical relationship is only exactly satisfied for a single instant. In fact, the problem is one of far greater complexity than that of ordinary disturbed rotation. The system is disturbed periodically, but the periodic time of the disturbance slowly increases, passing through a phase of equality to the free periodic time; the problem is to find the amplitude of the oscillations when they are at their maximum, and to find the mean configuration of the system some time before and some time after the maximum, when the oscillations are small. This problem does not seem to be soluble, unless we take into account the slow variation of the argument in the periodic disturbing term; and when the argument varies, the disturbing term is not strictly a simple time harmonic.

In the case of the viscous spheroid, the question would be further complicated by the fact that when the nutation becomes large, a new series of bodily tides is set up by the effects of inertia.

I have been unable to make a satisfactory examination of this problem, but as far as I have gone it appeared to me probable that the mean obliquity of the axis of the spheroid would not be affected by the passage of the system through a phase of large nutation; and although I cannot pretend to say how large the nutation might be, yet I consider it probable that the amplitude would not have time to increase to a very wide extent.*

* I believe that I shall be able to show in an investigation, as yet incomplete, that when this critical phase is reached, the plane of the lunar orbit is nearly coincident with the equator of the earth. As the amplitude of this nutation depends on the sine of the obliquity of the equator to the lunar orbit, it seems probable that the nutation would not become considerable.—June 30, 1879.

Throughout all the preceding investigations, the periodic inequalities have been neglected. Now a full development of the couples $\mathfrak{L}$, $\mathfrak{M}$, $\mathfrak{N}$, which are due to the tides, shows that there occur terms of speeds $n-2\Omega$, and $n-4\Omega$ in the first two, and of speeds 2Ω and 4Ω in the last. The terms in $n-2\Omega$ in $\mathfrak{L}$ and $\mathfrak{M}$ will clearly give rise to an increasing nutation at the critical point which we are considering, but they will be so very much smaller than those arising out of the attraction on the permanent equatorial protuberance that they may be neglected. The terms in $n-4\Omega$ are multiplied by very small quantities, and I think it may safely be assumed that the system would pass through the critical phase where $\frac{1}{2}\frac{n^3}{\mathfrak{g}}+n=4\Omega$ with sufficient rapidity to prevent the nutation becoming large.

If we were to go to higher orders of approximation in the disturbing forces, it is clear that we should meet with an infinite number of critical phases, but the coefficients representing the amplitudes of the resulting nutations would be multiplied by such small quantities that they may safely be neglected.

§ 18. *The initial condition of the earth and moon.**

It is now supposed that, when the earth's rotation has been tracked back to where it is equal to twice the moon's orbital motion, the obliquity to the plane of the lunar orbit has become zero. Then it is clear that, as long as there is any relative motion of the earth and moon, the tidal friction and reaction must continue to exist, and n and Ω must tend to an equality. The previous investigation shows also that for small viscosity, however nearly n approaches Ω, the position of zero obliquity is dynamically stable.

As n is approaching Ω, the changes must have taken place more and more slowly in time. For if the earth was a cooling spheroid, it is unreasonable to suppose that the process of becoming less stiff in consistency (which has hitherto been supposed to be taking place, as we go backwards in time) could ever have been reversed; and if it were not reversed, then the lunar tides must have lagged by less and less, as more and more time was given by the slow relative motion of the two bodies for the moon's attraction to have its full effect. Hence the effects of the sun's attraction must again become sensible, after passing through a phase of insensibility—a phase perhaps short in time, but fertile in changes in the system. I shall not here make the attempt to trace the reappearance of these solar terms.

It is, however, possible to make a rough investigation of what must have been the initial state from which the earth and moon started the course of development, which has been tracked back thus far. To do this, it is only necessary to consider the equation of conservation of moment of momentum.

* For further consideration of this subject, see a paper on the "Secular Effects of Tidal Friction," 'Proc. Roy. Soc.,' No. 197, 1879. The arithmetic of this section has been recomputed since the paper was presented.

When the obliquity is neglected, that equation may be written $\frac{n}{n_0}=1+\mu\left\{1-\left(\frac{\Omega_0}{\Omega}\right)^{\frac{1}{3}}\right\}$, and it is proposed to find what values of n would make n equal to Ω.

In the course of the above investigation four different starting points were taken, viz.: those at the beginning of each period of integration. There are objections to taking any one of these, to give the numerical values required for the solution of the above equation; for, on the one hand, the errors of each period accumulate on the next, and therefore it is advantageous to take one of the early periods; whilst, on the other hand, in the early periods the values of the quantities are affected by the sensibility of the solar terms, and by the obliquity of the ecliptic. The beginning of the fourth period was chosen, because by that time the solar terms had become insignificant. At that epoch I found $\log n_0=3{\cdot}84753$, when the present tropical year is the unit of time, and $\mu={\cdot}6659$, μ being the ratio of the orbital moment of momentum to the earth's moment of momentum; also $\log s=5{\cdot}39378-10$, s being a constant. Now put $x^3=n=\Omega$, and we have

$$x^4-(1+\mu)n_0x+\frac{1}{s}=0$$

Then substituting the numerical values,

$$x^4-11727x+40385=0$$

This equation has two real roots, one of which is nearly equal to $\sqrt[3]{11727}$, and the other to $40385\div11727$. By HORNER's method these roots are found to be 21·4320 and 3·4559 respectively. These are the two values of the cube root of the earth's rotation, for which the earth and moon move round as a rigid body.

The first gives a day of 5 hours 36 minutes, and the second a day of about 55½ m. s. days.

The latter is the state to which the earth and moon tend, under the influence of tidal friction (whether of oceanic or bodily tides) in the far distant future. For this case THOMSON and TAIT give a day of 48 of our present days;* the discrepancy between my value and theirs is explicable by the fact that they are considering a heterogeneous earth, whilst I treat a homogeneous one. Since on the hypothesis of heterogeneity the earth's moment of inertia is about $\frac{1}{3}Ma^2$, whilst on that of homogeneity it is $\frac{2}{5}Ma^2$, and since the $\frac{2}{5}$ which occurs in the quantity s enters by means of the expression for the earth's moment of inertia, it follows that in my solution μ has been taken too small in the proportion 5 : 6. Hence if we wish to consider the case of heterogeneity, we must solve the equation $x^4-12664x+48462=0$. The two roots of this equation are such that they give as the corresponding lengths of the day, 5 hours 16 minutes and 40·4 days respectively. The remaining discrepancy (between 40 and 48) is doubtless due in part

* 'Nat. Phil.,' § 276. They say:—"It is probable that the moon, in ancient times liquid or viscous in its outer layer or throughout, was thus brought to turn always the same face to the earth." In the new edition (1879) the ultimate effects of tidal friction are considered.

to the crude method of amending the solution, but also to the fact that they partly include the obliquity in one way, whilst I partly include it in another way, and I include a large part of the solar tidal friction whilst they neglect it. It is interesting to note that the larger root, which gives the shorter length of day, is but little affected by the consideration of the earth's heterogeneity.

With respect to the second solution (56 days), it must be remarked that the sun's tidal friction will go on lengthening the day even beyond this point, but then the lunar tides will again come into existence, and the lunar tidal friction will tend in part to counteract the solar. The tidal reaction will also be reversed, so that the moon will again approach the earth. Thus the effect of the sun is to make this a state of dynamical instability.

The first solution, where both the day and month are 5 hours 36 minutes long, is the one which is of interest in the present inquiry, for this is the initial state towards which the integration has been running back.

* This state of things is one of dynamical instability, as may be shown as follows :—

First consider the case where the sun does not exist. Suppose the earth to be rotating in about $5\frac{1}{2}$ hours, and the moon moving orbitally around it in a little less than that time. Then the motion of the moon relatively to the earth is consentaneous with the earth's rotation, and therefore the tidal friction, small though it be, tends to accelerate the earth's rotation ; the tidal reaction is such as to tend to retard the moon's linear velocity, and therefore increase her orbital angular velocity, and reduce her distance from the earth. The end will be that the moon falls into the earth.

This subject is graphically illustrated in a paper on the "Secular Effects of Tidal Friction," read before the Royal Society on June 19, 1879.

Secondly, take the case where the sun also exists, and suppose the system started in the same way as before. Now the motion of the earth relatively to the sun is rapid, and such that the solar tidal friction retards the earth's rotation ; whilst the lunar tidal friction is, as before, such as to accelerate the rotation.

Hence if the viscosity be very large the earth's rotation may be accelerated, but if it be not very large it will be retarded. The tidal reaction, which depends on the lunar tides alone, continues negative, and the moon approaches the earth as before. Thus after a short time the motion of the moon relatively to the earth is more rapid than in the previous case, whatever be the ratio between solar and lunar tidal friction. Hence in this case the moon will fall into the earth more rapidly than if the sun did not exist, and the dynamical instability is more marked.

If, however, the day were shorter than the month, the moon must continually recede from the earth, until it reaches the outer limit of a day of 56 m. s. days.

There is one circumstance which might perhaps decide that this should be the direction in which the equilibrium would break down ; for the earth was a cooling

* From here to the end of the section a good many alterations have been made since the paper was presented.—July 5, 1879.

body, and therefore probably a contracting one, and therefore its rotation would tend to increase. Of course this increase of rotation is partly counteracted by the solar tidal friction, but on the present theory, the mere existence of the moon seems to show that it was not more than counteracted, for if it had been so the moon must have been drawn into and confounded with the earth.

This month of 5 hours 36 minutes corresponds to a lunar distance of 2·52 earth's mean radii, or about 10,000 miles; the month of 5 hours 16 minutes corresponds to 2·39 earth's mean radii; so that in the case of the earth's homogeneity only 6,000 miles intervene between the moon's centre and the earth's surface, and even this distance would be reduced if we treated the earth as heterogeneous. This small distance seems to me to point to a break-up of the earth-moon mass into two bodies at a time when they were rotating in about 5 hours; for of course the precise figures given above cannot claim any great exactitude (see also Section 23).

It is a material circumstance in the conditions of the breaking-up of the earth into two bodies to consider what would have been the ellipticity of the earth's figure when rotating in $5\frac{1}{2}$ hours. Now the reciprocal of the ellipticity of a homogeneous fluid or viscous spheroid varies as the square of the period of rotation of the spheroid. The reciprocal of the ellipticity for a rotation in 24 hours is 232, and therefore the reciprocal of the ellipticity for a rotation in $5\frac{1}{2}$ hours is $(\frac{5\frac{1}{2}}{24})^2$ of $232=\frac{121}{2304}\times 232=12\cdot 2$.

Hence the ellipticity of the earth when rotating in $5\frac{1}{2}$ hours is $\frac{1}{12}$th.

The conditions of stability of a rotating mass of fluid are as yet unknown, but when we look at the planets Jupiter and Saturn, it is not easy to believe that an ellipticity of $\frac{1}{12}$th is sufficiently great to cause the break-up of the spheroid.

A homogeneous fluid spheroid of the same density as the earth has its greatest ellipticity compatible with equilibrium when rotating in 2 hours 24 minutes.*

The maximum ellipticity of all fluid spheroids of the same density is the same, and their periods of rotation multiplied by the square root of their densities is a function of the ellipticity only. Hence a spheroid, which rotates in 4 hours 48 minutes, will be in limiting equilibrium if its density is $(\frac{2\cdot 4}{4\cdot 8})^2$ or $\frac{1}{4}$ of that of the earth. If this latter spheroid had the same mass as the earth, its radius would be $\sqrt[3]{4}$ or 1·59 of that of the earth. If therefore the earth had a radius of 6,360 miles, and rotated in 4 hours 48 minutes, it would just have the maximum ellipticity compatible with equilibrium. It is, however, by no means certain that instability would not have set in long before this limiting ellipticity was reached.

In Part III. I shall refer to another possible cause of instability, which may perhaps be the cause of the break-up of the earth into two bodies.

It is easy to find the minimum time in which the system can have passed from this initial configuration, where the day and month are both $5\frac{1}{2}$ hours, down to the present

* PRATT's 'Fig. of Earth,' 2nd edition., Arts. 68 and 70.

condition. If we neglect the obliquity of the ecliptic, the equation (57) of tidal reaction, when adapted to the case of a viscous spheroid, becomes

$$\mu\frac{d\xi}{dt}=\frac{1}{2}\frac{\tau^2}{\mathfrak{g}n_0}\sin 4\epsilon_1$$

Now it is clear that the rate of tidal reaction can never be greater than when $\sin 4\epsilon_1=1$, when the lunar semi-diurnal tide lags by $22\frac{1}{2}°$. Then since $\tau=\frac{\tau_0}{\xi^6}$, we shall obtain the minimum time by integrating the equation

$$\frac{dt}{d\xi}=2\mu\frac{\mathfrak{g}n_0}{\tau_0^2}\xi^{12}$$

Whence

$$-t=\frac{2\mu}{13}\frac{\mathfrak{g}n_0}{\tau_0^2}(1-\xi^{13})$$

Now $\xi=\left(\frac{\Omega_0}{\Omega}\right)^{\frac{1}{3}}$, and we have found by the solution of the biquadratic that the initial condition is given by $\Omega^{\frac{1}{3}}=21\cdot4320$; also with the present value of the month $\Omega_0^{\frac{1}{3}}=4\cdot38$, the present year being in both cases the unit of time. Hence it follows that ξ is very nearly $\cdot2$, and ξ^{13} may be neglected compared with unity. Thus $-t=\frac{2\mu}{13}\frac{\mathfrak{g}n_0}{\tau_0^2}$.

Now $\mu=4\cdot007$ and $\frac{\mathfrak{g}n_0}{\tau_0^2}$ is 86,844,000 years.

Hence $-t=53{,}540{,}000$ years.

Thus we see that tidal reaction is competent to reduce the system from the initial state to the present state in something over 54 million years.

The rest of the paper is occupied with the consideration of a number of miscellaneous points, which it was not convenient to discuss earlier.

§ 19. *The change in the length of year.*

The effects of tidal reaction on the earth's orbit round the sun have been neglected; I shall now justify that neglect, and show by how much the length of the year may have been altered.

It is easy to show that the moment of momentum of the orbital motion of the moon and earth round their common centre of inertia is $\frac{C}{s\Omega^{\frac{1}{3}}}$, where C is the earth's moment of inertia, and $s=\frac{2}{5}\left[\left(\frac{a\nu}{g}\right)^2(1+\nu)\right]^{\frac{1}{3}}$.

The moment of momentum of the earth's rotation is obviously Cn. The normal to the lunar orbit is inclined to the earth's axis at an angle i. Hence the resultant moment of momentum of the moon and earth is

$$C\left\{n^2+\frac{1}{(s\Omega^{\frac{1}{3}})^2}+\frac{2n}{s\Omega^{\frac{1}{3}}}\cos i\right\}^{\frac{1}{2}}$$

The change in this quantity from one epoch to another is the amount of moment of momentum of the moon-earth system which has been destroyed by solar tidal friction. This destroyed moment of momentum reappears in the form of moment of momentum of the moon and earth in their orbital motion round the sun.

Now at the beginning of the integration of Section 17, that is to say at the present time, I find that when the present year is taken as the unit of time, the resultant moment of momentum of the moon and earth is 11369 C.

At the end of the third period of integration (after which the solar terms were neglected), and when the obliquity has become 15° 22′, I find the same quantity to be 11625 C.

Hence the loss of moment of momentum is 256 C., or 102·4 Ma^2.

Now at the present time the moment of momentum of the moon and earth in their orbit is $(M+m)\Omega_{,}c_{,}^2=Ma^2.\frac{1+\nu}{\nu}\left(\frac{c_{,}}{a}\right)^2\Omega_{,}$; $\frac{a}{c_{,}}$ is clearly the sun's parallax, and with the present unit of time $\Omega_{,}$ is 2π.

Hence the loss of moment of momentum is equal to the present moment of momentum of orbital motion multiplied by $\frac{102\cdot 4}{2\pi}\frac{\nu}{1+\nu}$ (sun's parallax)^2.

But the moment of momentum of the earth's and moon's orbital motion round the sun varies as $\Omega_{,}^{-\frac{1}{3}}$; hence the loss of moment of momentum corresponding to a change of $\Omega_{,}$ to $\Omega_{,}+\delta\Omega_{,}$ is the present moment of momentum multiplied by $\frac{1}{3}\frac{\delta\Omega_{,}}{\Omega_{,}}$, whence it is clear that

$$\frac{\delta\Omega_{,}}{\Omega_{,}}=3\frac{102\cdot 4}{2\pi}\cdot\frac{\nu}{1+\nu}\cdot\times\text{(sun's parallax)}^2.$$

But the shortening of the year is $\frac{\delta\Omega_{,}}{\Omega_{,}}$ of a year; taking therefore the sun's parallax as 8″·8, we find that at the end of the third period of integration the year was shorter than at present by

$$3\times\frac{102\cdot 4}{2\pi}\times\frac{82}{83}\times\left(\frac{8\cdot 8\pi}{648{,}000}\right)^2\times 365\cdot 25\times 86{,}400\text{ seconds,}$$

which will be found equal to 2·77 seconds.

Thus the solar tidal reaction had only the effect of lengthening the year by 2¾ seconds, since the epoch specified as the end of the third period of integration. The whole change in the length of year since the initial condition to which we traced back the moon would probably be very small indeed, but it is impossible to make this assertion positively, because, as observed above, the solar effects must have again become sensible, after passing through a period of insensibility.

§ 20. *Terms of the second order in the tide-generating potential.*

The whole of the previous investigation has been conducted on the hypothesis that the tide-generating potential, estimated per unit volume of the earth's mass, is $w\tau r^2(\cos^2 \text{PM}-\frac{1}{3})$,* but in fact this expression is only the first term of an infinite series. I shall now show what kind of quantities have been neglected by this treatment. According to the ordinary theory, the next term of the tide-generating potential is

$$V_2 = w\frac{m}{c}\left(\frac{r}{c}\right)^3(\tfrac{5}{2}\cos^3 \text{PM}-\tfrac{3}{2}\cos \text{PM})$$

Although for my own satisfaction I have completely developed the influence of this term in a similar way to that exhibited at the beginning of this paper, yet it does not seem worth while to give so long a piece of algebra; and I shall here confine myself to the consideration of the terms which will arise in the tidal friction from this term in the potential, when the obliquity is neglected. A comparison of the result with the value of the tidal friction, as already obtained, will afford the requisite information as to what has been neglected.

Now when the obliquity is put zero (see Plate 36, fig. 1),

$$\cos \text{PM} = \sin\theta \sin(\phi-\omega)$$

where ω is written for $n-\Omega$ for brevity. Then

$$\cos^3 \text{PM} = \tfrac{3}{4}\sin^3\theta\sin(\phi-\omega)-\tfrac{1}{4}\sin^3\theta\sin 3(\phi-\omega)$$

and

$$\cos^3 \text{PM}-\tfrac{3}{5}\cos \text{PM} = \tfrac{3}{20}\sin\theta(1-5\cos^2\theta)\sin(\phi-\omega)-\tfrac{1}{4}\sin^3\theta\sin 3(\phi-\omega).$$

Then since

$$w\frac{m}{c}\left(\frac{r}{c}\right)^3\frac{5}{2} = w\tau\frac{r^3}{c}\frac{5}{3}$$

therefore

$$V_2 \div w\frac{\tau}{c}r^3 = -\tfrac{5}{12}\sin^3\theta\sin 3(\phi-\omega)+\tfrac{1}{4}\sin\theta(1-5\cos^2\theta)\sin(\phi-\omega)$$

If $\sin 3(\phi-\omega)$ and $\sin(\phi-\omega)$ be expanded, we have V_2 in the desired form, viz.: a series of solid harmonics of the third degree, each multiplied by a simple time harmonic. Now if $wr^3 S_3 \cos(vt+\eta)$ be a tide-generating potential, estimated per unit volume of a homogeneous perfectly fluid spheroid of density w, S_3 being a surface harmonic of the third order, then the equilibrium tide due to this potential is given by $\sigma = \frac{7a^3}{4g}S_3\cos(vt+\eta)$, or $\frac{\sigma}{a} = \frac{7a}{10g}S_3\cos(vt+\eta)$. Hence just as in Section 2, the tide-

* See Section 1.

generating potential of the third order due to the moon will raise tides in the earth, when there is a frictional resistance to the internal motion, given by

$$\frac{\sigma}{a}=\frac{7}{10}\frac{\tau a}{\mathfrak{g}c}\left[-\tfrac{5}{12}F\sin^3\theta\sin 3(\phi-\omega+f)+\tfrac{1}{4}F'\sin\theta\,(1-5\cos^2\theta)\sin(\phi-\omega+f')\right]$$

Now σ is a surface harmonic of the third order, and therefore the potential of this layer of matter, at an external point whose coordinates are r, θ, ϕ, is

$$\frac{4}{7}\pi a w\left(\frac{a}{r}\right)^4\sigma=\frac{3}{7}\frac{Ma^2}{r^4}\sigma$$

Hence the moment about the earth's axis of the forces which the attraction of the distorted spheroid exercises on a particle of mass m, situated at r, θ, ϕ, is $\frac{3}{7}\frac{Mma}{r^4}\frac{d\sigma}{d\phi}$. Now if this mass be equal to that of the moon, and $r=c$, then $\frac{3}{7}\frac{Mma^2}{r^4}=\frac{2}{7}\frac{\tau}{c}Ma^2=\frac{5}{7}\frac{\tau}{c}\text{C}$, where, as before, C is the moment of inertia of the earth.

Hence the couple $\mathfrak{N}_2$, which the moon's attraction exercises on the earth, is given by $\mathfrak{N}_2=-\frac{5}{7}\frac{\tau}{c}\text{C}\frac{d\sigma}{d\phi}$, where after differentiation we put $\theta=\frac{\pi}{2}$ and $\phi=\frac{\pi}{2}+\omega$.

Now

$$-\frac{d\sigma}{d\phi}=\frac{7}{10}\frac{\tau}{\mathfrak{g}}\frac{a^2}{c}\cdot\left[\tfrac{5}{4}F\sin^3\theta\cos 3(\phi-\omega+f)-\tfrac{1}{4}F'\sin\theta(1-5\cos^2\theta)\cos(\phi-\omega+f')\right]$$

Hence

$$\frac{\mathfrak{N}_2}{\text{C}}\div\frac{1}{2}\frac{\tau^2}{\mathfrak{g}}\left(\frac{a}{c}\right)^2=\tfrac{5}{4}F\cos\left(\frac{3\pi}{2}+3f\right)-\tfrac{1}{4}F'\cos\left(\frac{\pi}{2}+f'\right)$$
$$=\tfrac{5}{4}F\sin 3f+\tfrac{1}{4}F'\sin f'$$

In the case of viscosity

$$F=\cos 3f,\ F'=\cos f'$$

Therefore

$$\frac{\mathfrak{N}_2}{\text{C}}=\left(\frac{a}{c}\right)^2\frac{\tau^2}{\mathfrak{g}}\left(\tfrac{5}{16}\sin 6f+\tfrac{1}{16}\sin 2f'\right)$$

Now if the obliquity had been neglected, the tidal friction $\mathfrak{N}_1$, due to the term of the first order in the tide-generating potential, would be given by $\frac{\mathfrak{N}_1}{\text{C}}=\frac{\tau^2}{\mathfrak{g}}\frac{1}{2}\sin 4\epsilon_1$.

Hence

$$\frac{\mathfrak{N}_2}{\mathfrak{N}_1}=\frac{1}{8}\left(\frac{a}{c}\right)^2\left(\frac{5\sin 6f+\sin 2f'}{\sin 4\epsilon_1}\right)$$

That is to say, this is the ratio of the terms neglected previously to those included.

Now according to the theory of viscous tides,*

$$\tan 3f = \frac{2\cdot 4^2+1}{3}\,\frac{(3\omega)}{gwa}\,v = \tfrac{22}{19}(3\omega)\left(\frac{19v}{2gwa}\right)$$

where v is the coefficient of viscosity.

But throughout the previous work we have written $\rho = \frac{2gwa}{19v}$.

Hence $\tan 3f = \frac{22}{19}\frac{3\omega}{\rho}$, and similarly $\tan f' = \frac{22}{19}\frac{\omega}{\rho}$.

Also $\tan 2\epsilon_1 = \frac{2\omega}{\rho}$.

I will now consider two cases :—

1st. Suppose the viscosity to be small, then f, f', ϵ_1 are all small, and

$$\frac{\sin 6f}{\sin 4\epsilon_1} = \frac{\tan 3f}{\tan 2\epsilon_1} = \tfrac{22}{19}\times\tfrac{3}{2},\quad \frac{\sin 2f'}{\sin 4\epsilon_1} = \frac{\tan f'}{\tan 2\epsilon_1} = \tfrac{22}{19}\times\tfrac{1}{2}$$

Therefore

$$\frac{\mathfrak{N}_2}{\mathfrak{N}_1} = \tfrac{22}{19}\left(\frac{a}{c}\right)^2$$

2nd. Suppose the viscosity very great, then $3f, f', 2\epsilon_1$ are very nearly equal to $\frac{\pi}{2}$, and $\tan\left(\frac{\pi}{2}-3f\right) = \frac{19}{22}\frac{\rho}{3\omega}$, $\tan\left(\frac{\pi}{2}-f'\right) = \frac{19}{22}\frac{\rho}{\omega}$, $\tan\left(\frac{\pi}{2}-2\epsilon\right) = \frac{\rho}{2\omega}$, so that we have approximately

$$\frac{\sin 6f}{\sin 4\epsilon_1} = \frac{\sin(\pi-6f)}{\sin(\pi-4\epsilon_1)} = \tfrac{19}{22}\times\tfrac{2}{3}$$

and similarly

$$\frac{\sin 2f'}{\sin 4\epsilon_1} = \tfrac{19}{22}\times 2$$

So that

$$\frac{\mathfrak{N}_2}{\mathfrak{N}_1} = \left(\frac{a}{c}\right)^2 \tfrac{1}{8}\times\tfrac{19}{22}\left(\tfrac{10}{3}+2\right) = \tfrac{19}{33}\left(\frac{a}{c}\right)^2$$

Hence it follows that the terms of the second order may bear a ratio to those of the first order lying between $\frac{22}{19}\left(\frac{a}{c}\right)^2$, or $1\cdot16\left(\frac{a}{c}\right)^2$, and $\frac{19}{33}\left(\frac{a}{c}\right)^2$, or $\cdot576\left(\frac{a}{c}\right)^2$.

Now at the end of the fourth period of integration in the solution of Section 15, $\frac{c}{a}$ or the moon's distance in earth's mean radii was 9; hence the terms of the second order in the equation of tidal friction must at that epoch lie in magnitude between $\frac{1}{70}$th and $\frac{1}{141}$st of those of the first order. It follows, therefore, that even at that stage, when the moon is comparatively near the earth, the effect of the tides of the second order (*i.e.*, of the third degree of harmonics) is insignificant, and the neglect of them is justified.

In the case of those terms of this order, which affect the obliquity, a very similar relationship to the terms of the lower order would be found to hold good.

* "Bodily Tides," &c., Phil. Trans., 1879, Part I., Section 5.

§ 21. *On certain other small terms.*

It will be well to advert to certain terms, the neglect of which might be suspected of vitiating my results.

According to the hypothesis of the plastic nature of the earth's mass, that body must have been a figure of equilibrium at every time throughout the series of changes which are to be followed out. In consequence of tidal friction the earth's rotation is diminishing, and therefore its ellipticity (which by the ordinary theory is $\frac{5}{4}\frac{n^2a}{g}$) is also diminishing; this change of figure might be supposed to exercise a material influence on the results, but I will now show that in one respect at least its effects are unimportant.

In a previous paper* I showed that, neglecting $\frac{C-A}{A}$ compared with unity, when the earth's figure changed symmetrically with respect to the axis of rotation,

$$\frac{di}{dt}=-\frac{\tau+\tau_{\prime}}{Cn^2}\sin i\cos i\frac{d}{dt}(C-A)$$

Now if e be the ellipticity of figure

$$C-A=\tfrac{2}{3}Ma^2e$$

So that

$$\frac{1}{C}\frac{d}{dt}(C-A)=\frac{de}{dt}=\frac{5}{2}\frac{na}{g}\frac{dn}{dt}=-\frac{n}{\mathfrak{g}}\frac{\mathfrak{N}}{C}$$

and therefore

$$\frac{di}{dt}=\frac{\tau+\tau_{\prime}}{\mathfrak{g}n}\sin i\cos i\frac{\mathfrak{N}}{C}$$

Now numerical calculation shows that at present $\frac{\tau+\tau_{\prime}}{\mathfrak{g}}=\frac{3\cdot04}{10^7}$, and since $\frac{\mathfrak{N}}{Cn}\sin i\cos i$ is of the same order of magnitude as $\frac{\mathfrak{L}}{Cn}, \frac{\mathfrak{M}}{Cn}$ (on which the changes of obliquity have been shown to depend), it follows that this term is fairly negligeable compared with those already included in the equations. As far as it goes, however, this term tends in the direction of increasing the obliquity with the time.†

* "On the Influence of Geological Changes," &c., Phil. Trans, Vol. 167, Part I., page 272, Section 8. The notation is changed, and the equation presented in a form suitable for the present purpose.

† In a paper in the 'Phil. Mag.,' March, 1877, I suggested that the obliquity might possibly be due to the contraction of the terrestrial nebula in cooling; I there neglected tidal friction and assumed the conservation of moment of momentum to hold good for the earth by itself, so that the ellipticity was continually increasing with the time. I did not at that time perceive that this increase of ellipticity was antagonistic to the effects of contraction. Though the work of that paper is correct, as I believe, yet the fundamental assumption is incorrect, and therefore the results are not worthy of attention.

[It will however appear, I believe, that this secular change of ellipticity of the earth's figure will exercise an important influence on the plane of the lunar orbit and thereby will affect the secular change in the obliquity of the ecliptic. The investigation of this point is however as yet incomplete.]*

The other small term which I shall consider arises out of the ordinary precession, together with the fact that the tide-generating force diminishes with the time on account of the tidal reaction on the moon.

The differential equations which give the ordinary precession are in effect (compare equations (26))

$$\frac{d\omega_1}{dt} = \tau\frac{C-A}{C}\sin i\cos i\sin n$$

$$\frac{d\omega_2}{dt} = -\tau\frac{C-A}{C}\sin i\cos i\cos n$$

and they give rise to no change of obliquity if τ be constant, but

$$\tau = \frac{\tau_0}{\xi^6} = \tau_0\left\{1 - 6\left(\frac{d\xi}{dt}\right)t\right\}$$

when t is small.

Also $\frac{C-A}{C} = e = \frac{5n^2a}{4g} = \frac{1}{2}\frac{n^2}{\mathfrak{g}}$. Hence as far as regards the change of obliquity the equations may be written

$$\frac{d\omega_1}{dt} = -\frac{3\tau_0 n^2}{\mathfrak{g}}\left(\frac{d\xi}{dt}\right)\sin i\cos i\ t\sin n$$

$$\frac{d\omega_2}{dt} = \frac{3\tau_0 n^2}{\mathfrak{g}}\left(\frac{d\xi}{dt}\right)\sin i\cos i\ t\cos n$$

Then if we regard all the quantities, except t, on the right-hand sides of these equations as constants and integrate, we have

$$\omega_1 = \frac{3\tau_0}{\mathfrak{g}}\left(\frac{d\xi}{dt}\right)\sin i\cos i\{nt\cos n - \sin n\}$$

$$\omega_2 = \frac{3\tau_0}{\mathfrak{g}}\left(\frac{d\xi}{dt}\right)\sin i\cos i\{nt\sin n + \cos n\}$$

And if these be substituted in the geometrical equations (1) we have

$$\frac{di}{dt} = \frac{3\tau_0}{\mathfrak{g}}\sin i\cos i\left(\frac{d\xi}{dt}\right)$$

* Added July 3, 1879.

Now by comparing this with the small term due to the secular change of figure of the earth, we see that it is fairly negligeable, being of the same order of magnitude as that term. As far as it goes, however, it tends to increase the obliquity of the ecliptic.

§ 22. *The change of obliquity and tidal friction due to an annular satellite.*

Conceive the ring to be rotating round the planet with an angular velocity Ω, let its radius be c, and its mass per unit length of its arc $\frac{m}{2\pi c}$, so that its mass is m. Let cl be the length of the arc measured from some point fixed in the ring up to the element $c\delta l$; and let Ωt be the longitude of the fixed point in the ring at the time t. Let δV be the tide-generating potential due to the element $\frac{m}{2\pi}\delta l$. Then we have by (5)

$$\delta V \div w r^2 \frac{3}{2c^3}\left(\frac{m}{2\pi}\delta l\right) = -(\xi^2-\eta^2)\Phi_l - 2\xi\eta\Phi'_l - \&c.$$

Where the suffixes to the functions indicate that $\Omega+l$ is to be written for Ω. Then integrating all round the ring from $l=0$ to $l=2\pi$ it is clear that

$$\frac{V}{w\tau r^2} = -p^2q^2\sin^2\theta\cos 2(\phi-n) + 2pq(p^2-q^2)\sin\theta\cos\theta\cos(\phi-n) + (\tfrac{1}{3}-\cos^2\theta)\tfrac{1}{2}(1-6p^2q^2)$$

which is the tide-generating potential of the ring.

Hence, as in Section 2, the form of the tidally-distorted spheroid is given by (9), save that E_1, E_2, E'_1, E'_2, E'' are all zero. Also, as in that section, the moments of the forces which the tidally-distorted spheroid exerts on the element of ring are $\frac{3}{5}\left(\frac{m}{2\pi}\delta l\right)\frac{Ma}{r^3}\left(\eta\frac{d\sigma}{d\zeta}-\zeta\frac{d\sigma}{d\eta}\right)$, &c., &c., where ξr, ηr, ζr are put equal to the rectangular coordinates of the element of ring, whose annular coordinate is l.

Now if x, y, z are the direction cosines of the element, equations (7) are simply modified by Ω being written $\Omega+l$. Hence the couples due to one element of ring may be found just as the whole couples were found before, and the integrals of the elementary couples from $l=0$ to 2π are the desired couples due to the whole ring. Now a little consideration shows that the results of this integration may be written down at once by putting E_1, E_2, E'_1, E'_2, E'' zero in (15), (16), and (21). Thus in order to determine the change of obliquity and the tidal friction due to an annular satellite, we have simply the expressions (33) and (34), save that $\tau\tau_1$ must be replaced by $\frac{1}{2}\tau^2$.

It thus appears that an annular satellite causes tidal friction in its planet, and that the obliquity of the planet's axis to the ring tends to diminish, but both these effects are evanescent with the obliquity. Since this ring only raises the tides which are called sidereal semi-diurnal and sidereal diurnal, and since we see by (57), Section 14, that tidal reaction is independent of those tides, it follows that there is no tangential force on the ring tending to accelerate its linear motion. If, however, the arc of the ring be not of uniform density, there is a slight tendency for the lighter parts to gain on the heavier, and the heavier parts become more remote from the planet than the lighter.

§ 23. *Double tidal reaction.*

Throughout the whole of this investigation the moon has been supposed to be merely an attractive particle, but there can be no doubt but that, if the earth was plastic, the moon was so also. To take a simple case, I shall now suppose that both the earth and moon are homogeneous viscous spheres revolving round their common centre of inertia, and that the moon is rotating on her own axis with an angular velocity ω, and that their axes are parallel and perpendicular to the plane of their orbit. Then the whole of the argument with respect to the earth as disturbed by the moon, may be transferred to the case of the moon as disturbed by the earth.

All symbols which apply to the moon will be distinguished from those which apply to the earth by an accent.

Then from (21) or (43) we have

$$\frac{\mathfrak{N}'}{C'}=\tfrac{1}{2}\frac{\tau'^2}{\mathfrak{g}'}\sin 4\epsilon'_1$$

and the equation which gives the lunar tidal friction is

$$\frac{d\omega}{dt}=-\tfrac{1}{2}\frac{\tau'^2}{\mathfrak{g}'}\sin 4\epsilon'_1. \qquad . \qquad (89)$$

Now

$$\tau'=\tfrac{3}{2}\frac{M}{c^3}=\nu\tau=\frac{wa^3}{w'a'^3}\tau$$

and

$$\mathfrak{g}'=\tfrac{2}{5}\frac{g'}{a'}=\frac{2g}{5a}\frac{w'}{w}=\frac{w'}{w}\mathfrak{g}$$

So that

$$\frac{\tau'^2}{\mathfrak{g}'}=\left(\frac{wa^2}{w'a'^2}\right)^3\frac{\tau^2}{\mathfrak{g}} \qquad (90)$$

Also

$$\frac{C'}{C}=\frac{w'a'^5}{wa^5}$$

and therefore

$$\frac{\mathfrak{N}'}{C}=\tfrac{1}{2}\frac{\tau^2}{\mathfrak{g}}\frac{w^2a}{w'^2a'}\sin 4\epsilon'_1$$

Now the force on the moon tangential to her orbit, results from a double tidal reaction. By the method employed in Section 14, the tangential force due to the earth's tides is

$$\mathrm{T}=\frac{\mathfrak{N}}{r}=\frac{\mathrm{C}}{2r}\frac{\tau^2}{\mathfrak{g}}\sin 4\epsilon_1$$

and similarly the tangential force due to the moon's tides is

$$\mathrm{T}'=\frac{\mathfrak{N}'}{r}=\frac{\mathrm{C}}{2r}\frac{\tau^2}{\mathfrak{g}}\frac{w^2a}{w'^2a'}\sin 4\epsilon'_1$$

and the whole tangential force is (T+T').

Hence following the argument of that section, the equation of tidal reaction becomes

$$\mu\frac{d\xi}{dt}=\tfrac{1}{2}\frac{\tau^2}{\mathfrak{g}n_0}\left[\sin 4\epsilon_1+\frac{w^2a}{w'^2a'}\sin 4\epsilon'_1\right]$$

Then taking the moon's apparent radius as 16′, and the ratio of the earth's mass to that of the moon as 82, we have $\frac{a}{a'}=3\cdot567$ and $\frac{w}{w'}=1\cdot806$ (so that taking w as $5\frac{1}{2}$, the specific gravity of the moon is 3), and hence $\frac{w^2a}{w'^2a'}=11\cdot64$.

At first sight it would appear from this that the effect of the tides in the moon was nearly twelve times as important as the effect of those in the earth, as far as concerns the influence on the moon's orbit, and hence it would seem that a grave oversight has been made in treating the moon as a simple attractive particle; a little consideration will show, however, that this is by no means the case.

Suppose that υ', υ are the coefficients of viscosity of the moon and earth respectively; then the only tides which exist in each body being those of which the speeds are $2(\omega-\Omega)$, $2(n-\Omega)$ in the moon and earth respectively,

$$\tan 2\epsilon'_1=\frac{19\upsilon'(\omega-\Omega)}{g'a'w'}\text{ and }\tan 2\epsilon_1=\frac{19\upsilon(n-\Omega)}{gaw}$$

But

$$g'a'w'=gaw\left(\frac{w'a'}{wa}\right)^2$$

and hence

$$\tan 2\epsilon'_1=\frac{\omega-\Omega}{n-\Omega}\frac{\upsilon'}{\upsilon}\left(\frac{wa}{w'a'}\right)^2\tan 2\epsilon_1$$

It will be found that $\left(\frac{wa}{w'a'}\right)^2=41\cdot10$. It is also almost certain that υ' must for a long time be greater than v, because the moon being a smaller body must have stiffened quicker than the earth. Hence unless $\omega-\Omega$ is very much less than $n-\Omega$, ϵ'_1 must be larger than ϵ_1. Therefore if in the early stages of development the earth had a small viscosity, it is probable that the effects of the moon's tides on her own orbit must have had a much more important influence than had the tides in the earth.

I shall now show, however, that this state of things must probably have had so short a duration as not to seriously affect the investigation of this paper. By (89) and (90) we have, as the equation which determines the rate of tidal friction reducing the moon's rotation round her axis,

$$\frac{d\omega}{dt}=-\tfrac{1}{2}\frac{\tau^2}{\mathfrak{g}}\left(\frac{wa^2}{w'a'^2}\right)^3\sin 4\epsilon'_1$$

Now $\left(\frac{wa^2}{w'a'^2}\right)^3=12{,}148$; and hence, for the same values of ϵ'_1 and ϵ_1, the moon's rotation round her axis is reduced 12,000 times as rapidly as that of the earth round its axis, and therefore in a very short period the moon's rotation round her axis must have been reduced to a sensible identity with the orbital motion. As ω becomes very nearly equal to Ω, $\sin 4\epsilon'_1$ becomes very small. Hence the term in the equation of tidal reaction dependent on the moon's own tides must have become rapidly evanescent. Now while this shows that the main body of our investigation is unaffected by the lunar tide, there is one slight modification of them to which it leads.

In Section 18 we traced back the moon to the initial condition, when her centre was 10,000 miles from the earth's centre. If lunar tidal friction had been included, this distance would have been increased; for the coefficient of x in the biquadratic (viz.: 11,727) would have to be diminished by $\frac{w'a'^5}{wa^5}(\omega-\omega_0)$. Now $\frac{w'a'^5}{wa^5}$ is very nearly $\frac{1}{1000}$th, and the unit of time being the year, it follows that we should have to suppose an enormously rapid primitive rotation of the moon round her axis, to make any sensible difference in the configuration of the two bodies when her centre of inertia moved as though rigidly connected with the earth's surface.

The supposition of two viscous globes moving orbitally round their common centre of inertia, and one having a congruent and the other an incongruent axial rotation, would lead to some very curious results.

§ 24. *Secular contraction of the earth.**

If the earth be contracting as it cools, it follows, from the principle of conservation of moment of momentum, that the angular velocity of rotation is being increased. Sir WILLIAM THOMSON has, however, shown that the contraction (which probably now only takes place in the superficial strata) cannot be sufficiently rapid to perceptibly counteract the influence of tidal friction at the present time.

* Rewritten in July, 1879.

The enormous height of the lunar mountains compared to those in the earth seems, however, to give some indications that a cooling celestial orb must contract by a perceptible fraction of its radius after it has consolidated.* Perhaps some of the contraction might be due to chemical combinations in the interior, when the heat had departed, so that the contraction might be deep-seated as well as superficial.

It will be well, therefore, to point out how this contraction will influence the initial condition to which we have traced back the earth and moon, when they were found rotating as parts of a rigid body in a little more than 5 hours.

Let C, C_0 be the moment of inertia of the earth at any time, and initially. Then the equation of conservation of moment of momentum becomes

$$\frac{Cn}{C_0 n_0}=1+\mu\left(1-\left(\frac{\Omega_0}{\Omega}\right)^{\frac{1}{3}}\right)$$

And the biquadratic of Section 18 which gives the initial configuration becomes

$$x^4-(1+\mu)\cdot\frac{C_0 n_0}{C}x+\frac{C_0}{Cs}=0$$

The required root of this equation is very nearly equal to $\left[(1+\mu)\frac{C_0 n_0}{C}\right]^{\frac{1}{3}}$. Now $x^3=\Omega$; hence Ω is nearly equal to $(1+\mu)\frac{C_0 n_0}{C}$. But in Section 18, when C was equal to C_0, it was nearly equal to $(1+\mu)n_0$. Therefore on the present hypothesis, the value

* Suppose a sphere of radius a to contract until its radius is $a+\delta a$, but that, its surface being incompressible, in doing so it throws up n conical mountains, the radius of whose bases is b, and their height h, and let b be large compared with h. The surface of such a cone is $\pi b\sqrt{h^2+b^2}=\pi(b^2+\frac{1}{2}h^2)$. Hence the excess of the surface of the cone above the area of the base is $\frac{1}{2}\pi h^2$, and $4\pi a^2=4\pi(a+\delta a)^2+\frac{1}{2}n\pi h^2$. Therefore $-\frac{\delta a}{a}=\frac{n}{16}\left(\frac{h}{a}\right)^2$.

Then suppose we have a second sphere of primitive radius a', which contracts and throws up the same number of mountains; then similarly $-\frac{\delta a'}{a'}=\frac{n}{16}\left(\frac{h'}{a'}\right)^2$ and $\frac{\delta a'}{a'}\div\frac{\delta a}{a}=\left(\frac{h'a}{ha'}\right)^2$. Now let these two spheres be the earth and moon. The height of the highest lunar mountain is 23,000 feet (GRANT's 'Physical Astron.,' p. 229), and the height of the highest terrestrial mountain is 29,000 feet; therefore we may take $\frac{h'}{h}=\frac{23}{29}$. Also $\frac{a'}{a}=\cdot 2729$ (HERSCHEL's 'Astron.,' Section 404). Therefore $\frac{ha'}{h'a}=\frac{29}{23}$ of $\cdot 2729=\cdot 344$, and $\left(\frac{ha'}{h'a}\right)^2=\cdot 1183$ or $\left(\frac{h'a}{ha'}\right)^2=8\cdot 45$. Hence $\frac{\delta a'}{a'}\div\frac{\delta a}{a}=8\frac{1}{2}$; whence it appears that, if both lunar and terrestrial mountains are due to the crumpling of the surfaces of those globes in contraction, the moon's radius has been diminished by about eight times as large a fraction as the earth's.

This is, no doubt, a very crude way of looking at the subject, because it entirely omits volcanic action from consideration, but it seems to justify the assertion that the moon has contracted much more than the earth, since both bodies solidified.

of Ω as given in that section must be multiplied by $\frac{C_0}{C}$; and the periodic time must be multiplied by $\frac{C}{C_0}$. But in this initial state C is greater than C_0; hence the periodic time when the two bodies move round as a rigid body is longer, and the moon is more distant from the earth, if the earth has sensibly contracted since this initial configuration.

If, then, the theory here developed of the history of the moon is the true one, as I believe it is, it follows that the earth cannot have contracted since this initial state by so much as to considerably diminish the effects of tidal friction, and it follows that Sir WILLIAM THOMSON's result as to the present unimportance of the contraction must have always been true.

If the moon once formed a part of the earth we should expect to trace the changes back until the two bodies were in actual contact. But it is obvious that the data at our disposal are not of sufficient accuracy, and the equations to be solved are so complicated, that it is not to be expected that we should find a closer accordance, than has been found, between the results of computation and the result to be expected, if the moon was really once a part of the earth.

It appears to me, therefore, that the present considerations only negative the hypothesis of any large contraction of the earth since the moon has existed.

PART III.

*Summary and discussion of results.**

The general object of the earlier or preparatory part of the paper is sufficiently explained in the introductory remarks.

The earth is treated as a homogeneous spheroid, and in what follows, except where otherwise expressly stated, the matter of which it is formed is supposed to be purely viscous. The word "earth" is thus an abbreviation of the expression "a homogeneous rotating viscous spheroid;" also wherever numerical values are given they are taken from the radius, mean density, mass, &c., of the earth.

The case is considered first of the action of one tide-raising body, namely, the moon. To simplify the problem the moon is supposed to move in a circular orbit in the ecliptic†—that plane being the average position of the lunar orbit with respect to the

* This part has been altered in accordance with the several additions and alterations occurring above. The results of subsequent investigations have modified the interpretation to be put on several of the results here obtained. I have, moreover, had the advantage of discussing several points with Sir WILLIAM THOMSON.—July 9, 1879.

† The effect of neglecting the eccentricity of the moon's orbit is, that we underestimate the efficiency of the tidal effects. Those effects vary as the inverse sixth power of r the radius vector, and if T be the

earth's axis. The case becomes enormously more complex if we suppose the moon to move in an inclined eccentric orbit with revolving nodes. The consideration of the secular changes in the inclination of the lunar orbit and of the eccentricity will form the subject of another investigation.

The expression for the moon's tide-generating potential is shown to consist of 13 simple tide-generating terms, and the physical meaning of this expansion is given in the note to Section 8. The physical causes represented by these 13 terms raise 13 simple tides in the earth, the heights and retardations of which depend on their speeds and on the coefficient of viscosity.

The 13 simple tides may be more easily represented both physically and analytically as seven tides, of which three are approximately semi-diurnal, three approximately diurnal, and one has a period equal to a half of the sidereal month, and is therefore called the fortnightly tide.

Then by an approximation which is sufficiently exact for a great part of the investigation, the semi-diurnal tides may be grouped together, and the diurnal ones also. Hence the earth may be regarded as distorted by two complex tides, namely, the semi-diurnal and diurnal, and one simple tide, namely, the fortnightly. The absolute heights and retardations of these three tides are expressed by six functions of their speeds and of the coefficient of viscosity (Sections 1 and 2).

When the form of the distorted spheroid is thus given, the couples about three axes fixed in the earth due to the attraction of the moon on the tidal protuberances are found. It must here be remarked that this attraction must in reality cause a tangential stress between the tidal protuberances and the true surface of the mean oblate spheroid. This tangential stress must cause a certain very small tangential flow,* and hence must ensue a very small diminution of the couples. The diminution of couple is here neglected, and the tidal spheroid is regarded as being instantaneously rigidly connected with the rotating spheroid. The full expression for the couples on the earth are long and complex, but since the nutations to which they give rise are exceedingly minute, they may be much abridged by the omission of all terms except such as can give rise to secular changes in the precession, the obliquity of the ecliptic, and the diurnal rotation. The terms retained represent that there are three couples independent of the time, the first of which tends to make the earth rotate about an axis in the equator which is always 90° from the nodes of the moon's orbit: this couple affects the obliquity to the ecliptic; second, there is a couple about an axis in

periodic time of the moon, the average value of $\frac{1}{r^6}$ is $\frac{1}{T}\int_0^T \frac{dt}{r^6}$. If c be the mean distance and e the eccentricity of the orbit, this integral will be found equal to $\frac{1}{c^6}\frac{1+3e^2+\frac{3}{8}e^4}{(1-e^2)^{\frac{9}{2}}}$. If the eccentricity be small the average value of $\frac{1}{r^6}$ is $\frac{1}{c^6}\left(1+\frac{15}{2}e^2\right)$; if e is $\frac{1}{20}$ this is $\frac{54}{53}$ of $\frac{1}{c^6}$. There are obviously forces tending to modify the eccentricity of the moon's orbit.

* See Part I. of the next paper.

the equator which is always coincident with the nodes: this affects the precession; third, there is a couple about the earth's axis of rotation, and this affects the length of the day (Sections 3, 4, and 5). All these couples vary as the fourth power of the moon's orbital angular velocity, or as the inverse sixth power of her distance.

These three couples give the alteration in the precession due to the tidal movement, the rate of increase of obliquity, and the rate at which the diurnal rotation is being diminished, or in other words the tidal friction. The change of obliquity is in reality due to tidal friction, but it is convenient to retain the term specially for the change of rotation alone.

It appears that if the bodily tides do not lag, which would be the case if the earth were perfectly fluid or perfectly elastic, then there is no alteration in the obliquity, nor any tidal friction (Section 7). The alteration in the precession is a very small fraction of the precession due to the earth considered as a rigid oblate spheroid. I have some doubts as to whether this result is properly applicable to the case of a perfectly fluid spheroid. At any rate, Sir WILLIAM THOMSON has stated, in agreement with this result, that a perfectly fluid spheroid has a precession scarcely differing from that of a perfectly rigid one. Moreover, the criterion which he gives of the negligeability of the additional terms in the precession in a closely analogous problem appears to be almost identical with that found by me (Section 7). I am not aware that the investigation on which his statement is founded has ever been published. The alteration in the precession being insignificant, no more reference will be made to it. This concludes the analytical investigation as far as concerns the effects on the disturbed spheroid, where there is only one disturbing body.

The sun is now (Section 8) introduced as a second disturbing body. Its independent effect on the earth may be determined at once by analogy with the effect of the moon. But the sun attracts the tides raised by the moon, and *vice versâ.* Now notwithstanding that the periods of the sun and moon about the earth have no common multiple, yet the interaction is such as to produce a secular alteration in the position of the earth's axis and in the angular velocity of its diurnal rotation. A physical explanation of this curious result is given in the note to Section 8. I have distinguished this from the separate effect of each disturbing body, as a combined effect.

The combined effects are represented by two terms in the tide-generating potential, one of which goes through its period in 12 sidereal hours, and the other in a sidereal day*; the latter being much more important than the former for moderate obliquities to the ecliptic. Both these terms vanish when the earth's axis is perpendicular to the plane of the orbit.

As far as concerns the combined effects, the disturbing bodies may be conceived to be

* These combined effects depend on the tides which are designated as K_1 and K_2 in the British Association's Report on Tides for 1872 and 1876, and which I have called the sidereal semi-diurnal and diurnal tides. For a general explanation of this result see the abstract of this paper in the 'Proceedings of the Royal Society,' No. 191, 1878.

replaced by two circular rings of matter coincident with their orbits and equal in mass to them respectively. The tidal friction due to these rings is insignificant compared with that arising separately from the sun and moon. But the diurnal combined effect has an important influence in affecting the rate of change of obliquity. The combined effects are such as to cause the obliquity of the ecliptic to diminish, whereas the separate effects on the whole make it increase—at least in general (see Section 22).

The relative importance of all the effects may be seen from an inspection of Table III., Section 15.

Section 11 contains a graphical analysis of the physical meaning of the equations, giving the rate of change of obliquity for various degrees of viscosity and obliquity.

Plate 36, figures 2 and 3, refer to the case where the disturbed planet is the earth, and the disturbing bodies the sun and moon.

This analysis gives some remarkable results as to the dynamical stability or instability of the system.

It will be here sufficient to state that, for moderate degrees of viscosity, the position of zero obliquity is unstable, but that there is a position of stability at a high obliquity. For large viscosities the position of zero obliquity becomes stable, and (except for a very close approximation to rigidity) there is an unstable position at a larger obliquity, and again a stable one at a still larger one.*

These positions of dynamical equilibrium do not rigorously deserve the name, since they are slowly shifting in consequence of the effects of tidal friction; they are rather positions in which the rate of change of obliquity becomes of a higher order of small quantities.

It appears that the degree of viscosity of the earth which at the present time would cause the obliquity of the ecliptic to increase most rapidly is such that the bodily semi-diurnal tide would be retarded by about 1 hour and 10 minutes; and the viscosity which would cause the obliquity to decrease most rapidly is such that the bodily semi-diurnal tide would be retarded by about $2\frac{3}{4}$ hours.

The former of these two viscosities was the one which I chose for subsequent numerical application, and for the consideration of secular changes in the system.

Plate 36, fig. 4 (Section 11), shows a similar analysis of the case where there is only one disturbing satellite, which moves orbitally with one-fifth of the velocity of rotation of the planet. This case differs from the preceding one in the fact that the position of zero obliquity is now unstable for all viscosities, and that there is always one other, and only one other position of equilibrium, and that is a stable one.

This shows that the fact that the *earth's* obliquity would diminish for large viscosity is due to the attraction of the sun on the lunar tides, and of the moon on the solar tides.

It is not shown by these figures, but it is the fact that if the motion of the satellite

* For a general explanation of some part of these results, see the abstract of this paper in the 'Proceedings of the Royal Society,' No. 191, 1878.

relatively to the planet be slow enough (viz.: the month less than twice the day), the obliquity will diminish.

This result, taken in conjunction with results given later with regard to the evolution of satellites, shows that the obliquity of a planet perturbed by a single satellite must rise from zero to a maximum and then decrease again to zero. If we regard the earth as a satellite of the moon, we see that this must have been the case with the moon.

Plate 36, fig. 5 (Section 12), contains a similar graphical analysis of the various values which may be assumed by the tidal friction. As might be expected, the tidal friction always tends to stop the planet's rotation, unless indeed the satellite's period is less than the planet's day, when the friction is reversed.

This completes the consideration of the effect on the earth, at any instant, of the attraction of the sun and moon on their tides; the next subject is to consider the reaction on the disturbing bodies.

Since the moon is tending to retard the earth's diurnal rotation, it is obvious that the earth must exercise a force on the moon tending to accelerate her linear velocity. The effect of this force is to cause her to recede from the earth and to decrease her orbital angular velocity. Hence tidal reaction causes a secular retardation of the moon's mean motion.

The tidal reaction on the sun is shown to have a comparatively small influence on the earth's orbit and is neglected (Sections 14 and 19).

The influence of tidal reaction on the lunar orbit is determined by finding the disturbing force on the moon tangential to her orbit, in terms of the couples which have been already found as perturbing the earth's rotation; and hence the tangential force is found in terms of the rate of tidal friction and of the rate of change of obliquity.

It appears that the non-periodic part of the force, on which the secular change in the moon's distance depends, involves the lunar tides alone.

By the consideration of the effects of the perturbing force on the moon's motion, an equation is found which gives the rate of increase of the square root of the moon's distance, in terms of the heights and retardations of the several lunar tides (Section 14).

Besides the interaction of the two bodies which affects the moon's mean motion, there is another part which affects the plane of the lunar orbit; but this latter effect is less important than the former, and in the present paper is neglected, since the moon is throughout supposed to remain in the ecliptic. The investigation of the subject will however, lead to interesting results, since a complete solution of the problem of the obliquity of the ecliptic cannot be attained without a simultaneous tracing of the secular changes in the plane of the lunar orbit.

It appears that the influence of the tides, here called slow semi-diurnal and slow diurnal, is to increase the moon's distance from the earth, whilst the influence of the fast semi-diurnal, fast diurnal, and fortnightly tide tends to diminish the moon's distance; also the sidereal semi-diurnal and diurnal tides exercise no effects in this respect. The two tides which tend to increase the moon's distance are much larger than the others, so that the moon in general tends to recede from the earth. The increase of distance is, of course, accompanied by an increase of the moon's periodic time, and hence there is in general a true secular retardation of the moon's motion. But this change is accompanied by a retardation of the earth's diurnal rotation, and a terrestrial observer, taking the earth as his clock, would conceive that the angular velocity of an ideal moon, which was undisturbed by tidal reaction, was undergoing a secular acceleration. The apparent acceleration of the ideal undisturbed moon must considerably exceed the true retardation of the real disturbed moon, and the difference between these two will give an apparent acceleration.

It is thus possible to give an equation connecting the apparent acceleration of the moon's motion and the heights and retardations of the several bodily tides in the earth.

Now there is at the present time an unexplained secular acceleration of the moon of about 4″ per century, and therefore if we attribute the whole of this to the action of the bodily tides in the earth, instead of to the action of ocean tides, as was done by ADAMS and DELAUNAY, we get a numerical relation which must govern the actual heights and retardations of the bodily tides in the earth at the present time.

This equation involves the six constants expressive of the heights and retardations of the three bodily tides, and which are determined by the physical constitution of the earth. No further advance can therefore be made without some theory of the earth's nature. Two theories are considered.

First, that the earth is purely viscous. The result shows that the earth is either nearly fluid—which we know it is not—or exceedingly nearly rigid. The only traces which we should ever be likely to find of such a high degree of viscosity would be in the fortnightly ocean tide; and even here the influence would be scarcely perceptible, for its height would be ·992 of its theoretical amount according to the equilibrium theory, whilst the time of high water would be only accelerated by six hours and a half.

It is interesting to note that the indications of a fortnightly ocean tide, as deduced from tidal observations, are exceedingly uncertain, as is shown in a preceding paper,* where I have made a comparison of the heights and phases of such small fortnightly tides as have hitherto been observed. And now (July, 1879) Sir WILLIAM THOMSON has informed me that he thinks it very possible that the effects of the earth's rotation may be such as to prevent our trusting to the equilibrium theory to give even approximately the height of the fortnightly tide. He has recently read a paper on this subject before the Royal Society of Edinburgh.

With the degree of viscosity of the earth, which gives the observed amount of secular acceleration to the moon, it appears that the moon is subject to such a true secular retardation that at the end of a century she is 3″·1 behind the place in her orbit which she would have occupied if it were not for the tidal reaction, whilst the earth, considered as a clock, is losing 13 seconds in the same time. This rate of retardation of the earth

* See the Appendix to my paper on the "Bodily Tides," &c., Phil. Trans., Part I., 1879.

is such that an observer taking the earth as his clock would conceive a moon, which was undisturbed by tidal reaction, to be 7″·1 in advance of her place at the end of a century. But the actual moon is 3″·1 behind her true place, and thus our observer would suppose the moon to be in advance 7·1—3·1 or 4″ at the end of the century. Lastly, the obliquity of the ecliptic is diminishing at the rate of 1° in 500 million years.

The other hypothesis considered is that the earth is very nearly perfectly elastic. In this case the semi-diurnal and diurnal tides do not lag perceptibly, and the whole of the reaction is thrown on to the fortnightly tide, and moreover there is no perceptible tidal frictional couple about the earth's axis of rotation. From this follows the remarkable conclusion that the moon may be undergoing a true secular acceleration of motion of something less than 3″·5 per century, whilst the length of day may remain almost unaffected. Under these circumstances the obliquity of the ecliptic must be diminishing at the rate of 1° in something like 130 million years.

This supposition leads to such curious results, that I investigated what state of things we should arrive at if we look back for a very long period, and I found that 700 million years ago the obliquity might have been 5° greater than at present, whilst the month would only be a little less than a day longer. The suppositions on which these results are based are such that they *necessarily* give results more striking than would be physically possible.

The enormous lapse of time which has to be postulated renders it in the highest degree improbable that more than a very small change in this direction has been taking place, and moreover the action of the ocean tides has been entirely omitted from consideration.

The results of these two hypotheses show what fundamentally different interpretations may be put to the phenomenon of the secular acceleration of the moon.

Sir WILLIAM THOMSON also has drawn attention to another disturbing cause in the fall of meteoric dust on to the earth.*

Under these circumstances, I cannot think that any estimate having any pretension to accuracy can be made as to the present rate of tidal friction.

Since the obliquity of the ecliptic, the diurnal rotation of the earth, and the moon's distance change, the whole system is in a state of flux; and the next question to be considered is to determine the state of things which existed a very long time ago (Part II.). This involved the integration of three simultaneous differential equations; the mathematical difficulties were, however, so great, that it was found impracticable to obtain a general analytical solution. I therefore had to confine myself to a numerical solution adapted to the case of the earth, sun, and moon, for one particular degree of viscosity of the earth. The particular viscosity was such that, with the present values of the day and month, the time of the lunar semi-diurnal tide was retarded by 1 hour and 10 minutes; the greatest possible lagging of this tide is

* 'Glasgow Geological Society,' Vol. III. Address "On Geological Time."

3 hours, and therefore this must be regarded as a very moderate degree of viscosity. It was chosen because initially it makes the rate of change of obliquity a maximum, and although it is not that degree of viscosity which will make all the changes proceed with the greatest possible rapidity, yet it is sufficiently near that value to enable us to estimate very well the smallest time which can possibly have elapsed in the history of the earth, if changes of the kind found really have taken place. This estimate of time is confirmed by a second method, which will be referred to later.

The changes were tracked backwards in time from the present epoch, and for convenience of diction I shall also reverse the form of speech—*e.g.*, a true loss of energy as the time increases will be spoken of as a gain of energy as we look backwards.

I shall not enter at all into the mathematical difficulties of the problem, but shall proceed at once to comment on the series of tables at the end of Section 15, which give the results of the solution.

The whole process, as traced backwards, exhibits a gain of kinetic energy to the system (of which more presently), accompanied by a transference of moment of momentum from that of orbital motion of the moon and earth to that of rotation of the earth. The last column but one of Table IV. exhibits the fall of the ratio of the two moments of momentum from 4·01 down to ·44. The whole moment of momentum of the moon-earth system rises slightly, because of solar tidal friction. The change is investigated in Section 19.

Looked at in detail, we see the day, month, and obliquity all diminishing, and the changes proceeding at a rapidly increasing rate, so that an amount of change which at the beginning required many millions of years, at the end only requires as many thousands. The reason of this is that the moon's distance diminishes with great rapidity; and as the effects vary as the square of the tide-generating force, they vary as the inverse sixth power of the moon's distance, or, in physical language, the height of the tides increases with great rapidity, and so also does the moon's attraction. But there is a counteracting principle, which to some extent makes the changes proceed slower. It is obvious that a disturbing body will not have time to raise such high tides in a rapidly rotating spheroid as in one which rotates slowly. As the earth's rotation increases, the lagging of the tides increases. The first column of Table I. shows the angle by which the crest of the lunar semi-diurnal tide precedes the moon; we see that the angle is almost doubled at the end of the series of changes, as traced backwards. It is not quite so easy to give a physical meaning to the other columns, although it might be done. In fact, as the rotation increases, the effect of each tide rises to a maximum, and then dies away; the tides of longer period reach their maximum effect much more slowly than the ones of short period. At the point where I have found it convenient to stop the solution (see Table IV.), the semi-diurnal effect has passed its maximum, the diurnal tide has just come to give its maximum effect, whilst the fortnightly tide has not nearly risen to that point.

As the lunar effects increase in importance (when we look backwards), the relative value of the solar effects decreases rapidly, because the solar tidal reaction leaves the earth's orbit sensibly unaffected (see Section 19), and thus the solar effects remain nearly constant, whilst the lunar effects have largely increased. The relative value of the several tidal effects is exhibited in Tables II. and III.

Table IV. exhibits the length of day decreasing to a little more than a quarter of its present value, whilst the obliquity diminishes through 9°. But the length of the month is the element which changes to the most startling extent, for it actually falls to $\frac{1}{17}$th of its primitive value.

It is particularly important to notice that all the changes might have taken place in 57 million years; and this is far within the time which physicists admit that the earth and moon may have existed. It is easy to find a great many *veræ causæ* for changes in the planetary system; but it is in general correspondingly hard to show that they are competent to produce any marked effects, without exorbitant demands on the efficiency of the causes and on lapse of time.

It is a question of great interest to geologists to determine whether any part of these changes could have taken place during geological history. It seems to me that this question must be decided by whether or not a globe, such as has been considered, could have afforded a solid surface for animal life, and whether it might present a superficial appearance such as we know it. These questions must, I think, be answered in the affirmative, for the following reasons.

The coefficient of viscosity of the spheroid with which the previous solution deals is given by the formula $\frac{wa}{19n}\tan 35°$ (see Section 11, (40)), when gravitation units of force are used. This, when turned into numbers, shows that $2{\cdot}055\times10^7$ grams weight are required to impart unit shear to a cubic centimeter block of the substance in 24 hours, or 2,055 kilogs. per square centimeter acting tangentially on the upper face of a slab one centimeter thick for 24 hours, would displace the upper surface through a millimeter relatively to the lower, which is held fixed. In British units this becomes,—13½ tons to the square inch, acting for 24 hours on a slab an inch thick, displaces the upper surface relatively to the lower through one-tenth of an inch. It is obvious that such a substance as this would be called a solid in ordinary parlance, and in the tidal problem this must be regarded as a rather small viscosity.

It seems to me, then, that we have only got to postulate that the upper and cool surface of the earth presents such a difference from the interior that it yields with extreme slowness, if at all, to the weight of continents and mountains, to admit the possibility that the globe on which we live may be like that here treated of. If, therefore, astronomical facts should confirm the argument that the world has really gone through changes of the kind here investigated, I can see no adequate reason for assuming that the whole process was pre-geological. Under these circumstances it must be admitted that the obliquity to the ecliptic is now probably slowly decreasing;

that a long time ago it was perhaps a degree greater than at present, and that it was then nearly stationary for another long time, and that in still earlier times it was considerably less.*

The violent changes which some geologists seem to require in geologically recent times would still, I think, not follow from the theory of the earth's viscosity.

According to the present hypothesis (and for the moment looking forward in time), the moon-earth system is, from a dynamical point of view, continually losing energy from the internal tidal friction. One part of this energy turns into potential energy of the moon's position relatively to the earth, and the rest developes heat in the interior of the earth. Section 16 contains the investigation of the amount which has turned to heat between any two epochs. The heat is estimated by the number of degrees Fahrenheit, which the lost energy would be sufficient to raise the temperature of the whole earth's mass, if it were all applied at once, and if the earth had the specific heat of iron.

The last column of Table IV., Section 15, gives the numerical results, and it appears therefrom that, during the 57 million years embraced by the solution, the energy lost suffices to heat the whole earth's mass 1760° Fahr.

It would appear at first sight that this large amount of heat, generated internally, must seriously interfere with the accuracy of Sir WILLIAM THOMSON's investigation of the secular cooling of the earth ;† but a further consideration of the subject in the next paper will show that this cannot be the case.

There are other consequences of interest to geologists which flow from the present hypothesis. As we look at the whole series of changes from the remote past, the ellipticity of figure of the earth must have been continually diminishing, and thus the polar regions must have been ever rising and the equatorial ones falling; but, as the ocean always followed these changes, they might quite well have left no geological traces.

The tides must have been very much more frequent and larger, and accordingly the rate of oceanic denudation much accelerated.

The more rapid alternations of day and night‡ would probably lead to more sudden and violent storms, and the increased rotation of the earth would augment the violence of the trade winds, which in their turn would affect oceanic currents.

Thus there would result an acceleration of geological action.

The problem, of which the solution has just been discussed, deals with a spheroid of

* In my paper "On the Effects of Geological Changes on the Earth's Axis," Phil. Trans. 1877, p. 271, I arrived at the conclusion that the obliquity had been unchanged throughout geological history. That result was obtained on the hypothesis of the earth's rigidity, except as regards geological upheavals. The result at which I now arrive affords a warning that every conclusion must always be read along with the postulates on which it is based.

† 'Nat. Phil.,' Appendix.

‡ At the point where the solution stops there are just 1,300 of the sidereal days of that time in the year, instead of 366 as at present.

constant viscosity; but there is every reason to believe that the earth is a cooling body, and has stiffened as it cooled. We therefore have to deal with a spheroid whose viscosity diminishes as we look backwards.

A second solution is accordingly given (Section 17) where the viscosity is variable; no definite law of diminution of viscosity is assumed, however, but it is merely supposed that the viscosity always remains small from a tidal point of view. This solution gives no indication of the time which may have elapsed, and differs chiefly from the preceding one in the fact that the change in the obliquity is rather greater for a given amount of change in the moon's distance.

There is not much to say about it here, because the two solutions follow closely parallel lines as far as the place where the former one left off.

The first solution was not carried further, because as the month approximates in length to the day, the three semi-diurnal tides cease to be of nearly equal frequencies, and so likewise do the three diurnal tides; hence the assumption on which the solution was founded, as to their approximately equal speeds, ceases to be sufficiently accurate.

In this second solution all the seven tides are throughout distinguished from one another. At about the stage where the previous solution stops the solar terms have become relatively unimportant, and are dropped out. It appears that (still looking backwards in time) the obliquity will only continue to diminish a little more beyond the point it had reached when the previous method had become inapplicable. For when the month has become equal to twice the day, there is no change of obliquity; and for yet smaller values of the month the change is the other way.

This shows that for small viscosity of the planet the position of zero obliquity is dynamically stable for values of the month which are less than twice the day, while for greater values it is unstable; and the same appears to be true for very large viscosity of the planet (see the foot-note on p. 500).

If the integration be carried back as far as the critical point of relationship between the day and month, it appears that the whole change of obliquity since the beginning is $9\frac{1}{2}°$.

The interesting question then arises—Does the hypothesis of the earth's viscosity afford a complete explanation of the obliquity of the ecliptic? It does not seem at present possible to give any very conclusive answer to this question; for the problem which has been solved differs in many respects from the true problem of the earth.

The most important difference from the truth is in the neglect of the secular changes of the plane of the lunar orbit; and I now (September, 1879) see reason to believe that that neglect will make a material difference in the results given for the obliquity at the end of the third and fourth periods of integration in both solutions. It will not, therefore, be possible to discuss this point adequately at present; but it will be well to refer to some other points in which our hypothesis must differ from reality.

I do not see that the heterogeneity of density and viscosity would make any very material difference in the solution, because both the change of obliquity and the tidal friction would be affected *pari passû*, and therefore the change of obliquity for a given amount of change in the day would not be much altered.

Although the effects of the contraction of the earth in cooling would be certainly such as to render the changes more rapid in time, yet as the tidal friction would be somewhat counteracted, the critical point where the month is equal to twice the day would be reached when the moon was further from the earth than in my problem. I think, however, that there is reason to believe that the whole amount of contraction of the earth, since the moon has existed, has not been large (Section 24).

There is one thing which might exercise a considerable influence favourable to change of obliquity. We are in almost complete ignorance of the behaviour of semi-solids under very great pressures, such as must exist in the earth, and there is no reason to suppose that the amount of relative displacement is simply proportional to the stress and the time of its action. Suppose, then, that the displacement varied as some other function of the time, then clearly the relative importance of the several tides might be much altered.

Now, the great obstacle to a large change of obliquity is the diurnal combined effect (see Table IV., Section 15); and so any change in the law of viscosity which allowed a relatively greater influence to the semi-diurnal tides would cause a greater change of obliquity, and this without much affecting the tidal friction and reaction. Such a law seems quite within the bounds of possibility. The special hypothesis, however, of elastico-viscosity, used in the previous paper, makes the other way, and allows greater influence to the tides of long period than to those of short. This was exemplified where it was shown that the tidal reaction might depend principally on the fortnightly tide.

The whole investigation is based on a theory of tides in which the effects of inertia are neglected. Now it will be shown in Part III. of the next paper that the effect of inertia will be to make the crest of the tidal spheroid lag more for a given height of tide than results from the theory founded on the neglect of inertia. An analysis of the effect produced on the present results, by the modification of the theory of tides introduced by inertia, is given in the next paper.

On the whole, we can only say at present that it seems probable that a part of the obliquity of the ecliptic may be referred to the causes here considered; but a complete discussion of the subject must be deferred to a future occasion, when the secular changes in the plane of the lunar orbit will be treated.

The question of the obliquity is now set on one side, and it is supposed that when the moon has reached the critical point (where the month is twice the day) the obliquity to the plane of the lunar orbit was zero. In the more remote past the obliquity had no tendency to alter, except under the influence of certain nutations, which are referred to at the end of Section 17.

The manner in which the moon's periodic time approximates to the day is an inducement to speculate as to the limiting or initial condition from which the earth and moon started their course of development.

So long as there is any relative motion of the two bodies there must be tidal friction, and therefore the moon's period must continue to approach the day. It would be a problem of extreme complication to track the changes in detail to their end, and fortunately it is not necessary to do so.

The principle of conservation of moment of momentum, which has been used throughout in tracing the parallel changes in the moon and earth, affords the means of leaping at once to the conclusion (Section 18). The equation expressive of that principle involves the moon's orbital angular velocity and the earth's diurnal rotation as its two variables; and it is only necessary to equate one to the other to obtain an equation, which will give the desired information.

As we are now supposed to be transported back to the initial state, I shall henceforth speak of time in the ordinary way; there is no longer any convenience in speaking of the past as the future, and *vice versâ.*

The equation above referred to has two solutions, one of which indicates that tidal friction has done its work, and the other that it is just about to begin. Of the first I shall here say no more, but refer the reader to Section 18.

The second solution indicates that the moon (considered as an attractive particle) moves round the earth as though it were rigidly fixed thereto in 5 hours 36 minutes. This is a state of dynamical instability; for if the month is a little shorter than the day, the moon will approach the earth, and ultimately fall into it; but if the day is a little shorter than the month, the moon will continually recede from the earth, and pass through the series of changes which were traced backwards.

Since the earth is a cooling and contracting body, it is likely that its rotation would increase, and therefore the dynamical equilibrium would be more likely to break down in the latter than the former way.

The continuous solution of the problem is taken up at the point where the moon has receded from the earth so far that her period is twice that of the earth's rotation.

I have calculated that the heat generated in the interior of the earth in the course of the lengthening of the day from 5 hours 36 minutes to 23 hours 56 minutes would be sufficient, if applied all at once, to heat the whole earth's mass about 3000° Fahr., supposing the earth to have the specific heat of iron (see Section 16).

A rough calculation shows that the minimum time in which the moon can have passed from the state where it had a period of 5 hours 36 minutes to the present state, is 54 million years, and this confirms the previous estimates of time.

This periodic time of the moon corresponds to an interval of only 6,000 miles between the earth's surface and the moon's centre. If the earth had been treated as heterogeneous, this distance, and with it the common periodic time both of moon and earth, would be still further diminished.

These results point strongly to the conclusion that, if the moon and earth were ever molten viscous masses, then they once formed parts of a common mass.

We are thus led at once to the inquiry as to how and why the planet broke up.

The conditions of stability of rotating masses of fluid are unfortunately unknown, and it is therefore impossible to do more than speculate on the subject.

The most obvious explanation is similar to that given in LAPLACE's nebular hypothesis, namely, that the planet being partly or wholly fluid, contracted, and thus rotated faster and faster until the ellipticity became so great that the equilibrium was unstable, and then an equatorial ring separated itself, and the ring finally conglomerated into a satellite. This theory, however, presents an important difference from the nebular hypothesis, in as far as that the ring was not left behind 240,000 miles away from the earth, when the planet was a rare gas, but that it was shed only 4,000 or 5,000 miles from the present surface of the earth, when the planet was perhaps partly solid and partly fluid.

This view is to some extent confirmed by the ring of Saturn, which would thus be a satellite in the course of formation.

It appears to me, however, that there is a good deal of difficulty in the acceptance of this view, when it is considered along with the numerical results of the previous investigation.

At the moment when the ring separated from the planet it must have had the same linear velocity as the surface of the planet; and it appears from Section 22 that such a ring would not tend to expand from tidal reaction, unless its density varied in different parts. Thus we should hardly expect the distance from the earth of the chain of meteorites to have increased much, until it had agglomerated to a considerable extent. It follows, therefore, that we ought to be able to trace back the moon's path, until she was nearly in contact with the earth's surface, and was always opposite the same face of the earth. Now this is exactly what has been done in the previous investigation. But there is one more condition to be satisfied, namely, that the common speed of rotation of the two bodies should be so great that the equilibrium of the rotating spheroid should be unstable. Although we do not know what is the limiting angular velocity of a rotating spheroid consistent with stability, yet it seems improbable that a rotation in a little over 5 hours, with an ellipticity of one-twelfth would render the system unstable.

Now notwithstanding that the data of the problem to be solved are to some extent uncertain, and notwithstanding the imperfection of the solution of the problem here given, yet it hardly seems likely that better data and a more perfect solution would largely affect the result, so as to make the common period of revolution of the two bodies in the initial configuration very much less than 5 hours.* Moreover we obtain no help from the hypothesis that the earth has considerably contracted since the shedding of the satellite, but rather the reverse; for it appears from Section 24 that if the earth has contracted, then the common period of revolution of the two bodies in the

* This is illustrated by my paper on "The Secular Effects of Tidal Friction," 'Proc. Roy. Soc.,' No. 197, 1879, where it appears that the "line of momentum" does not cut the "curve of rigidity" at a very small angle, so that a small error in the data would not make a very large one in the solution.

initial configuration must have been slower, and the moon more distant from the earth. This slower revolution would correspond with a smaller ellipticity, and thus the system would probably be less nearly unstable.

The following appears to me at least a possible cause of instability of the spheroid when rotating in about 5 hours. Sir WILLIAM THOMSON has shown that a fluid spheroid of the same mean density as the earth would perform a complete gravitational oscillation in 1 hour 34 minutes. The speed of oscillation varies as the square root of the density, hence it follows that a less dense spheroid would oscillate more slowly, and therefore a spheroid of the same mean density as the earth, but consisting of a denser nucleus and a rarer surface, would probably oscillate in a longer time than 1 hour 34 minutes. It seems to be quite possible that two complete gravitational oscillations of the earth in its primitive state might occupy 4 or 5 hours. But if this were the case, then the solar semi-diurnal tide would have very nearly the same period as the free oscillation of the spheroid, and accordingly the solar tides would be of enormous height.

Does it not then seem possible that, if the rotation were fast enough to bring the spheroid into anything near the unstable condition, then the large solar tides might rupture the body into two or more parts? In this case one would conjecture that it would not be a ring which would detach itself.

It seems highly probable that the moon once did rotate more rapidly round her own axis than in her orbit, and if she was formed out of the fusion together of a ring of meteorites, this rotation would necessarily result.

In Section 23 it is shown that the tidal friction due to the earth's action on the moon must have been enormous, and it must necessarily have soon brought her to present the same face constantly to the earth. This explanation was, I believe, first given by HELMHOLTZ. In the process, the inclination of her axis to the plane of her orbit must have rapidly increased, and then, as she rotated more and more slowly, must have slowly diminished again. Her present aspect is thus in strict accordance with the results of the purely theoretical investigation.

It would perhaps be premature to undertake a complete review of the planetary system, so as to see how far the ideas here developed accord with it. Although many facts which could be adduced seem favourable to their acceptance, I will only refer to two. The satellites of Mars appear to me a most remarkable confirmation of these views. Their extreme minuteness has prevented them from being subject to any perceptible tidal reaction, just as the minuteness of the earth compared with the sun has prevented the earth's orbit from being perceptibly influenced (see Section 19); they thus remain as a standing memorial of the primitive periodic time of Mars round his axis. Mars, on the other hand, has been subjected to solar tidal friction. This case, however, deserves to be submitted to numerical calculation.

The other case is that of Uranus, and this appears to be somewhat unfavourable to the theory; for on account of the supposed adverse revolution of the satellites, and of the high inclinations of their orbits, it is not easy to believe that they could have arisen from a planet which ever rotated about an axis at all nearly perpendicular to the ecliptic.

The system of planets revolving round the sun present so strong a resemblance to the systems of satellites revolving round the planets, that we are almost compelled to believe that their modes of development have been somewhat alike. But in applying the present theory to explain the orbits of the planets, we are met by the great difficulty that the tidal reaction due to solar tides in the planet is exceedingly slow in its influence; and not much help is got by supposing the tides in the sun to react on the planet. Thus enormous periods of time would have to be postulated for the evolution.

If, however, this theory should be found to explain the greater part of the configurations of the satellites round the planets, it would hardly be logical to refuse it some amount of applicability to the planets. We should then have to suppose that before the birth of the satellites the planets occupied very much larger volumes, and possessed much more moment of momentum than they do now. If they did so, we should not expect to trace back the positions of the axes of the planets to the state when they were perpendicular to the ecliptic, as ought to be the case if the action of the satellites, and of the sun after their birth, is alone concerned.

Whatever may be thought of the theory of the viscosity of the earth, and of the large speculations to which it has given rise, the fact remains that nearly all the effects which have been attributed to the action of bodily tides would also follow, though probably at a somewhat less rapid rate, from the influence of oceanic tides on a rigid nucleus. The effect of oceanic tidal friction on the obliquity of the ecliptic has already been considered by Mr. STONE, in the only paper on the subject which I have yet seen.* His argument is based on what I conceive to be an incorrect assumption as to the nature of the tidal frictional couple, and he neglects tidal reaction; he finds that the effects would be quite insignificant. This result would, I think, be modified by a more satisfactory assumption.

* Ast. Soc. Monthly Notices, March 8, 1867.

Fig. 1.

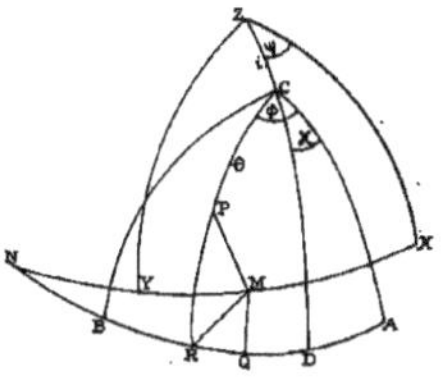

Fig. 2.

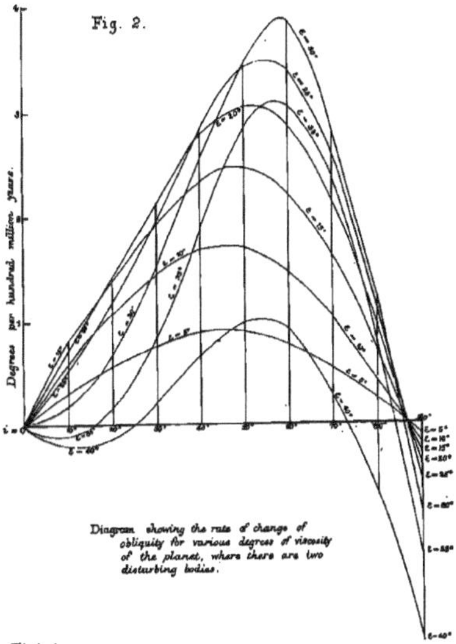

Diagram showing the rate of change of obliquity for various degrees of viscosity of the planet, where there are two disturbing bodies.

Fig. 3.

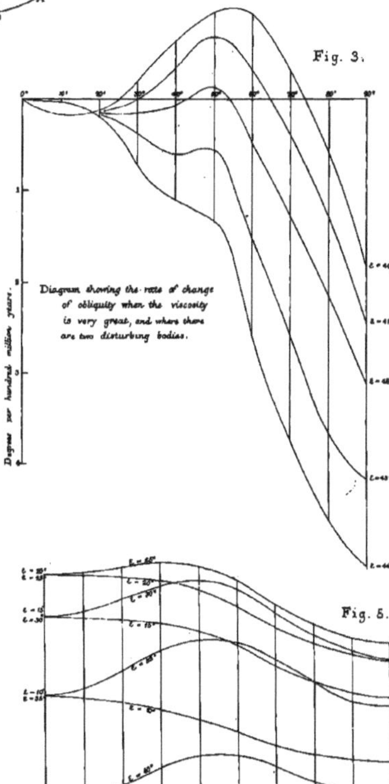

Diagram showing the rate of change of obliquity when the viscosity is very great, and where there are two disturbing bodies.

Fig. 4.

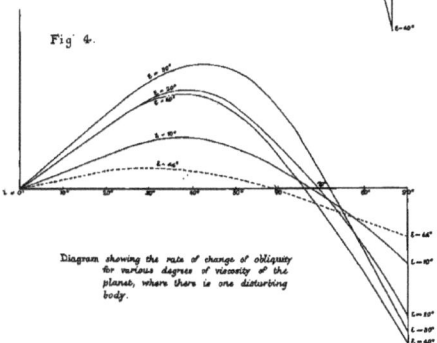

Diagram showing the rate of change of obliquity for various degrees of viscosity of the planet, where there is one disturbing body.

Fig. 5.

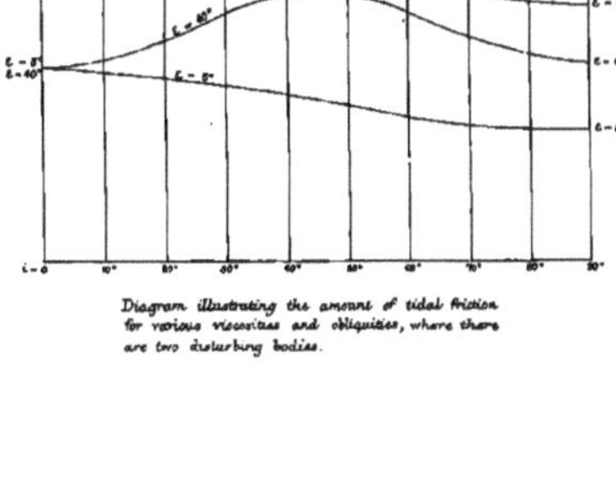

Diagram illustrating the amount of tidal friction for various viscosities and obliquities, where there are two disturbing bodies.

www.ingramcontent.com/pod-product-compliance
Lightning Source LLC
Chambersburg PA
CBHW022104210326
41519CB00056B/1200
9783337213985